Supply Chain Integration for Sustainable Advantages

Yongyi Shou · Mingu Kang · Young Won Park

Supply Chain Integration for Sustainable Advantages

Yongyi Shou
School of Management
Zhejiang University
Hangzhou, China

Mingu Kang
School of Management
Zhejiang University
Hangzhou, China

Young Won Park
Faculty of Economics
Saitama University
Saitama, Japan

ISBN 978-981-16-9331-1 ISBN 978-981-16-9332-8 (eBook)
https://doi.org/10.1007/978-981-16-9332-8

© The Editor(s) (if applicable) and The Author(s), under exclusive license to Springer Nature
Singapore Pte Ltd. 2022
This work is subject to copyright. All rights are solely and exclusively licensed by the Publisher, whether
the whole or part of the material is concerned, specifically the rights of translation, reprinting, reuse
of illustrations, recitation, broadcasting, reproduction on microfilms or in any other physical way, and
transmission or information storage and retrieval, electronic adaptation, computer software, or by similar
or dissimilar methodology now known or hereafter developed.
The use of general descriptive names, registered names, trademarks, service marks, etc. in this publication
does not imply, even in the absence of a specific statement, that such names are exempt from the relevant
protective laws and regulations and therefore free for general use.
The publisher, the authors and the editors are safe to assume that the advice and information in this book
are believed to be true and accurate at the date of publication. Neither the publisher nor the authors or
the editors give a warranty, expressed or implied, with respect to the material contained herein or for any
errors or omissions that may have been made. The publisher remains neutral with regard to jurisdictional
claims in published maps and institutional affiliations.

This Springer imprint is published by the registered company Springer Nature Singapore Pte Ltd.
The registered company address is: 152 Beach Road, #21-01/04 Gateway East, Singapore 189721,
Singapore

Acknowledgments

This book is impossible without the support of our colleagues and students. We would like to take this opportunity to express our sincere gratitude and appreciation.

First of all, we are grateful to our co-authors who have worked with us to publish journal articles on supply chain integration. We would like to extend our gratitude to the following scholars: Dr. Baofeng Huo at Tianjin University, Dr. Ying Li at Shandong University, Dr. Wenjin Hu at ETH Zurich, Dr. Ma Ga (Mark) Yang at West Chester University, and Min Tian and Lingjia Li at Zhejiang University.

We thank the postgraduate students at Zhejiang University who helped to prepare the book chapters. Ziwei Yang reviewed the relevant literature on supply chain integration and drafted Chapter 2. Xinyu Zhao contributed to preparing the final chapter. Xinyu Zhao, Xueshu Shan, Chang Wu, and Xinyi Fan helped to revise and proofread the book chapters.

Last but not least, we are thankful to the International Manufacturing Strategy Survey (IMSS) project team. The empirical studies in this book are all based on the data set from the sixth round IMSS. Thanks to the high-quality data set, we are able to conduct the studies presented in this book.

Contents

1 Introduction 1
   1.1 Supply Chain Integration 1
   1.2 Data Set 3
   1.3 An Overview 5
   References 6

2 A Systematic Literature Review of Supply Chain Integration 9
   2.1 Introduction 9
   2.2 Method 11
   2.3 Descriptive Analysis 12
      2.3.1 Publication Distributions 12
      2.3.2 Research Methods 13
   2.4 Thematic Analysis 15
      2.4.1 Theories in SCI Research 15
      2.4.2 Dimensions of SCI 18
      2.4.3 SCI and Performance 19
   2.5 Conclusions 23
   References 24

3 Product Complexity, Variety and Supply Chain Integration 31
   3.1 Introduction 31
   3.2 Literature Review and Hypotheses Development 33
      3.2.1 Product–Process Fit 33
      3.2.2 Product Complexity and SCI 35
      3.2.3 Product Variety and SCI 37
   3.3 Method 38
      3.3.1 Data 38
      3.3.2 Measures 39
      3.3.3 Reliability and Validity 39
   3.4 Results 41
   3.5 Discussion 43
      3.5.1 Findings 43

	3.5.2	Theoretical Implications	44
3.6	Conclusions		44
References			45

4 Production Systems and Supply Chain Integration 49
- 4.1 Introduction 49
- 4.2 Theoretical Background and Hypotheses Development 51
  - 4.2.1 SCI and Operational Performance 51
  - 4.2.2 The Contingency Effects of Production System 52
- 4.3 Method 54
  - 4.3.1 Data 54
  - 4.3.2 Measures 55
  - 4.3.3 Reliability and Validity 56
- 4.4 Results 58
- 4.5 Discussion 61
  - 4.5.1 Findings 61
  - 4.5.2 Theoretical Implications 61
- 4.6 Conclusions 64
- References 64

5 Enablers of Supply Chain Integration: A Socio-Technical System Perspective 67
- 5.1 Introduction 67
- 5.2 Literature Review and Hypotheses Development 70
  - 5.2.1 Socio-Technical System Perspective 70
  - 5.2.2 Human Resources 71
  - 5.2.3 Digital Manufacturing Technology 72
  - 5.2.4 Effects of HR on SCI 73
  - 5.2.5 Effects of DMT on SCI 74
  - 5.2.6 The Moderating Effect of Competition 75
- 5.3 Method 76
  - 5.3.1 Data 76
  - 5.3.2 Measures 77
  - 5.3.3 Reliability and Validity 77
- 5.4 Results 79
- 5.5 Discussion 82
  - 5.5.1 Findings 82
  - 5.5.2 Theoretical Implications 83
- 5.6 Conclusions 84
- References 85

6 Risk Management of Manufacturing Multinational Corporations: The Effects of Supply Chain Integration 91
- 6.1 Introduction 91
- 6.2 Literature Review and Hypotheses Development 94
  - 6.2.1 Organizational Information Processing Theory 94

| | Contents | ix |

6.2.2	SCRM and Operational Performance	95
6.2.3	The Moderating Effect of IAD	108
6.2.4	The Counteracting Effect of SCI	109

6.3 Method 111
    6.3.1 Data 111
    6.3.2 Measures 111
    6.3.3 Reliability and Validity 113
6.4 Results 113
6.5 Discussion 120
    6.5.1 Findings 120
    6.5.2 Theoretical Implications 120
6.6 Conclusions 121
References 122

7 Supply Chain Integration and Sustainability: The Supply Chain Learning Perspective 129
7.1 Introduction 129
7.2 Literature Review and Hypotheses Development 131
    7.2.1 Supply Chain Learning 131
    7.2.2 Sustainability Management Practices 133
    7.2.3 SCI and SMPs 134
    7.2.4 SMPs and Sustainability Performance 135
7.3 Method 136
    7.3.1 Data 136
    7.3.2 Measures 137
    7.3.3 Reliability and Validity 137
7.4 Results 139
    7.4.1 Structural Model Results 139
    7.4.2 Additional Analyses 140
7.5 Discussion 141
    7.5.1 Findings 141
    7.5.2 Theoretical Implications 142
7.6 Conclusions 143
References 143

8 Conclusions and Implications 149
8.1 A Brief Summary 149
8.2 Theoretical Implications 150
8.3 Managerial Implications 151
References 154

Appendix: Survey Questions 155

Chapter 1
Introduction

Abstract Supply chain integration (SCI) has been an important management practice for the firm's competitive advantages in the recent decades. Firms aim to integrate supply chain activities not only across functional departments but also cross organizational boundaries. In this book, we present a number of studies which investigate the antecedents and outcomes of SCI. Data from the sixth round International Manufacturing Strategy Survey (IMSS) project were used to empirically test the research models. The data set is introduced in detail, including its distribution, reliability, validity and measurement equivalence. This book is intended for contributing to the theory and practice about SCI.

Keywords Supply chain integration · Competitive advantage · IMSS

1.1 Supply Chain Integration

A supply chain is a network of interlinked external suppliers, manufacturing plants, and distribution channels organized to provide products and services for customers (Bidhandi et al., 2009; Park et al., 2012). The supply chain secures raw materials, transforms them into finished goods, and distributes them to customers through a network system (Park et al., 2009). Supply chain management needs to assess and satisfy both the current and future customers (end-users) (Al-Mudimigh et al., 2004). Focusing on relations between manufacturers and suppliers, intense global competition and complex customer expectations have led suppliers and focal manufacturers to collaborate and fulfill multiple customer requirements including cost, quality, innovativeness, delivery speed, dependability, and flexibility (Boyer & Lewis, 2002; Flynn & Flynn, 2004; Zhao et al., 2008). Collaborative network capabilities satisfy complex customer requirements that either manufacturer or suppliers alone may not satisfy (Bowersox et al., 1999; Park & Hong, 2019). Global supply chain networks are also characterized by a basic common goal and a highly integrated operational

---

This chapter is co-authored by Mingu Kang, Young Won Park and Yongyi Shou.

© The Author(s), under exclusive license to Springer Nature Singapore Pte Ltd. 2022
Y. Shou et al., *Supply Chain Integration for Sustainable Advantages*,
https://doi.org/10.1007/978-981-16-9332-8_1

system, combined with a complex decentralized decisional integration of different functions within a company and external integration with suppliers (Park & Hong, 2017; Zhao et al., 2011).

Accordingly, supply chain integration (SCI) between internal functions within a company and external collaborating partners has been a crucial point. Also, SCI as the strategic collaboration of both intra- and inter-organization processes has received increasing research attention as a practical way of achieving organizational goals (Ellegaard & Koch, 2012; He et al., 2014; Huang et al., 2014; Huo et al., 2013; Kull et al., 2019; Lai et al., 2012; Srinivasan & Swink, 2015; Tarifa-Fernandez & De Burgos-Jiménez, 2017). A crucial element of the global business activities of multinational corporations (MNCs) is the efficient design and operation of complex supply chain networks. Therefore, SCI has complex dimensions including information flows, cross-functional collaboration, and inter-organizational arrangements (Narasimhan & Kim, 2002; Park et al., 2009). Furthermore, SCI is recognized as an important factor to attain superior supply chain performance, as it offers a host of competitive advantages, including complete transparency from suppliers to customers.

Through SCI, global firms can attain sustainable competitive advantages instead of a mere source of cost reduction (Calle et al., 2015). In particular, during turbulent times, configuration, collaboration, and coordination complexities of the supply chain have been significant variables (Abdelkafi et al., 2011; Park et al., 2015). Thus, SCI has been increasingly becoming one of the most important competitive strategies used by modern enterprises. Organizations must design and integrate their supply chains to align the contingencies of the environment, strategy, and technology for survival and success (Hung et al., 2010). Because the main aim of supply chain management is to integrate various stakeholders including suppliers to satisfy market demand, selection and integration of key stakeholders such as suppliers and distributors in supply chain networks play important roles in establishing an effective supply chain as well as maintaining sustainable competitive advantages (Jia & Rutherford, 2010; Lee & Kim, 2008; Lin & Ho, 2009).

Given the strategic importance of integration, this book presents several research models that address the enabling factors and critical roles of SCI. First, three research models in this book investigate the antecedents of SCI at three different levels: product characteristics (product complexity and variety), production system configurations (one-of-a-kind, batch, and mass production), and socio-technical system factors (specifically, human resources and digital manufacturing technologies). The product complexity and variety cause manufacturing firms to cultivate SCI capability to mitigate transactional hazards and facilitate knowledge transfer (Shou et al., 2017). The fit between the distinct production system and external SCI dimensions offers deep insights into the nature of SCI (Shou et al., 2018). The socio-technical system in a manufacturing firm, including human resources and digital manufacturing technologies, enables the firm to promote SCI by facilitating information sharing, communication, and collaboration (Tian et al., 2021).

1.1 Supply Chain Integration

Second, the other two research models in this book examine the critical roles of SCI in operations management, including risk management and sustainability management practices. In other words, the research models not only confirm SCI as a vital enabler for operational performance in line with previous studies but also further investigates the roles of SCI in the above-mentioned areas which have not been sufficiently examined and deserve more attention. Integration with suppliers and customers helps a manufacturing firm to mitigate the risk caused by geographical dispersion in a global manufacturing network (Hu et al., 2020). Moreover, SCI also enhances intra- and inter-organizational sustainability management practices, thereby contributing to the firm's sustainability performance (Kang et al., 2018).

1.2 Data Set

To empirically examine the proposed research models, we specifically used the data collected from the sixth round of the International Manufacturing Strategy Survey (IMSS) project. The IMSS is a research project initially conducted in 1992, 1996, 2001, 2005, 2009, and 2013–2014 by a global network of universities and business schools. The IMSS network has generated a series of databases of manufacturing strategies, practices, and performance, contributing to the scientific community in general, and the area of operations and supply chain management in particular. The IMSS research group developed a standard English questionnaire, which was then translated into the local language by research coordinators using a reliable method (either double parallel translation or back-translation).

The IMSS questionnaire includes three sections. The first section seeks to obtain general information about the business unit, such as firm size, industry, competitive strategy, and business performance; the second section refers to the manufacturing practices and performance of the plant; and the third section pertains to supply chain practices. Eligible companies included in the IMSS project are manufacturing firms with more than 50 employees. Only companies with an ISIC Rev. 4 code between 25 and 30 were surveyed, using a random sampling approach. Specifically, the types of industry included: (25) manufacture of fabricated metal products, excluding machinery and equipment; (26) manufacture of computer, electronic and optical products; (27) manufacture of electrical equipment; (28) manufacture of machinery and equipment not elsewhere classified; (29) manufacture of motor vehicles, trailers and semi-trailers; and (30) manufacture of other transport equipment. The major survey respondents comprised production, operations, supply chain, or plant managers in manufacturing firms.

The six-round IMSS sample consists of data from 931 manufacturing companies from 22 countries and regions. These countries and regions were selected to represent the major industrial regions of the world, including America, Asia, and Europe. They also included both developed and developing countries, thus ensuring the representativeness of the sample. Especially, the data set collected from the sixth round

IMSS contains three dimensions of SCI (i.e., internal integration, supplier integration, and customer integration), thereby providing useful opportunities to examine the strategic role of SCI. The survey questions for SCI and other constructs used in this book are reported in the Appendix of this book.

The assessment and control of non-response bias are critical for the reliability and validity of survey-based research (Wagner & Kemmerling, 2010). The sixth round of the IMSS survey covered 22 countries and regions. The research coordinator in each country had to use a joint protocol to check non-response bias concerning the collected data (Demeter et al., 2016; Kauppi et al., 2016). Local researchers were permitted to adopt one of two approaches for testing the non-response bias. If a secondary data set of the firms in the surveyed country was available, local researchers could test for significant differences between respondents and non-respondents in terms of size, industry, sales, or proprietary structure. Otherwise, non-response bias was checked by using questionnaire items, such as size, industry, market share, and return on sales. No evidence of non-response bias was reported by local researchers.

Since the IMSS data were collected from single respondents, common method bias is a potential problem. The survey adopted some procedural remedies to mitigate potential bias, while the studies in this book also used statistical remedies (Podsakoff et al., 2003) to check whether common method bias significantly impacted the outcome. First, the variables used in the studies in this book were presented using separate scales in different sections of the questionnaire. Second, the IMSS survey promised to fully protect respondent anonymity through a clear statement on the cover page of the questionnaire. This procedure facilitated respondents' honest answers to the questions. In checking the significance of potential common method bias, we used Harman's single-factor model (Hu & Bentler, 1999) or a marker variable, which was theoretically unrelated to other variables. In the correlation between the marker and other variables, the insignificant coefficients indicate that common method bias is not a concern in the studies of this book (Williams et al., 2010).

Since the IMSS data were collected from 22 countries and regions, measurement equivalence across these countries and regions should be checked to ensure the validity of the data. We evaluated the measurement equivalence in terms of calibration, translation, and metric equivalence (Mullen, 1995). Calibration equivalence can be ensured since standardized Likert scales were used in the IMSS questionnaire to measure the constructs. Moreover, translation equivalence of all the survey items is guaranteed since the development of the IMSS questionnaire followed the careful translation guidelines and went through rigorous translation/back-translation processes (Kauppi et al., 2016). Metric equivalence is "the equivalence in the scoring process or the way respondents in different countries answer the same question" (Mullen, 1995). We applied multi-group confirmatory factor analysis (CFA) to assess the metric equivalence (Rungtusanatham et al., 2008). Specifically, we compared the goodness of fit of two models to assess metric equivalence. The first model is a fully constrained model with all factor loadings of each item being constrained to be equal across America, Asia, and Europe. The second model is an unconstrained model with all factor loadings of each item being freely estimated across continents. Cheung and Rensvold (2002) recommended the change in CFI (ΔCFI) value between

1.2 Data Set

the constrained and baseline models as an indicator of metric equivalence, which is insensitive to sample size and model parameters. The models in this book have small values of ΔCFI under the recommended threshold (e.g., Hu et al., 2020; Shou et al., 2017, 2018). Thus, metric equivalence is confirmed.

1.3 An Overview

The book is structured in eight chapters. This chapter introduces the important roles of SCI in this current business environment, the data set used in the analysis of research models that were presented in this book, and ends with a brief overview of the book's chapters.

Chapter 2 presents a systematic literature review on SCI. In recent decades, SCI has increasingly become one of the most popular topics in the areas of supply chain management. As a result, there has emerged a substantial body of research addressing the SCI issue. As SCI is a fundamental driver of manufacturing firms' competitiveness in a highly networked and complicated business environment, the previous studies of SCI must be well understood. Based on the systematic literature review, we offer a few recommendations about future research in this area.

Chapter 3 examines the antecedents of SCI at the product level. Specifically, the purposes of this chapter are to investigate the relationship between product-level characteristics (i.e., product complexity and product variety) and the three dimensions of SCI (i.e., internal, supplier, and customer integration). The results of this chapter extend knowledge regarding the product design–supply chain interface and also provide valuable insights into the ways that firms enhance SCI under conditions of a high level of product complexity and variety.

Chapter 4 focuses on external SCI (i.e., supplier integration and customer integration) and studies the roles of production system in determining the performance effects of SCI. Specifically, this chapter highlights the contingency effects of internal production systems (i.e., one-of-a-kind production, batch production, and mass production) on the relationship between external SCI and operational performance. The empirical findings suggest guidelines of how external SCI can be matched with different configurations of production systems to achieve the desired quality, flexibility, delivery, or cost performance.

Chapter 5, drawing on the socio-technical system (STS) perspective, investigates the effects of human resource and digital manufacturing technology (DMT) on SCI in the context of competition. In other words, this chapter identifies two important enablers including DMT (a technical factor) and human resource (a social factor) that a manufacturing firm can use to promote the three dimensions of SCI. Moreover, this chapter examines a contingent role of competition (an environmental factor) that can influence the effectiveness of human resource and DMT.

Chapter 6 looks at the ways in which SCI plays a contingent role in supply chain risk management (SCRM) practices of manufacturing MNCs. This chapter examines the SCRM–operational performance relationship and sheds light on how a manufacturing MNC's international asset dispersion negatively influences this relationship and how SCI can attenuate this negative moderating effect of international asset dispersion. The results of this chapter provide insights into how manufacturing MNCs can implement SCRM more effectively by leveraging appropriate external SCI.

Chapter 7 explores the role of SCI in promoting intra-organizational and inter-organizational sustainability management practices and sustainability performance (i.e., environmental performance, social performance, and economic performance). By incorporating SCI into the sustainability area beyond operations management, this chapter provides a new perspective on supply chain management and sustainability research.

Lastly, Chap. 8 provides conclusions of this book.

References

Abdelkafi, N., Pero, M., Blecker, T., & Sianesi, A. (2011). NPD-SCM alignment in mass customization. In F. Fogliatto & G. da Silveira (Eds.), *Mass customization*. Springer Series in Advanced Manufacturing. Springer. https://doi.org/10.1007/978-1-84996-489-0_4

Al-Mudimigh, A. S., Zairi, M., & Ahmed, A. M. M. (2004). Extending the concept of supply chain: The effective management of value chains. *International Journal of Production Economics, 87*(3), 309–320.

Bidhandi, H. M., Yusuff, R. M., Ahmad, M. M. H. M., & Bakar, M. R. A. (2009). Development of a new approach for deterministic supply chain network design. *European Journal of Operational Research, 198*(1), 121–128.

Bowersox, D. J., Closs, D. J., & Stank, T. P. (1999). *21st century logistics: Making supply chain integration a reality*. Council of Supply Chain Management Professionals.

Boyer, K. K., & Lewis, M. W. (2002). Competitive priorities: Investigating the need for trade-offs in operations strategy. *Production and Operations Management, 11*(1), 9–20.

Calle, A. D. L., Alvarez, E., & Freije, I. (2015). Supply chain integration, a key strategic capability for improving product and service value propositions: Empirical evidence. *International Journal of Engineering Management and Economics, 5*(1–2), 89–103.

Cheung, G. W., & Rensvold, R. B. (2002). Evaluating goodness-of-fit indexes for testing measurement invariance. *Structural Equation Modeling: A Multidisciplinary Journal, 9*(2), 233–255.

Demeter, K., Szász, L., & Rácz, B.-G. (2016). The impact of subsidiaries' internal and external integration on operational performance. *International Journal of Production Economics, 182*, 73–85.

Ellegaard, C., & Koch, C. (2012). The effects of low internal integration between purchasing and operations on suppliers' resource mobilization. *Journal of Purchasing and Supply Management, 18*(3), 148–158.

Flynn, B. B., & Flynn, E. J. (2004). An exploratory study of the nature of cumulative capabilities. *Journal of Operations Management, 22*(5), 439–457.

He, Y., Keung Lai, K., Sun, H., & Chen, Y. (2014). The impact of supplier integration on customer integration and new product performance: The mediating role of manufacturing flexibility under trust theory. *International Journal of Production Economics, 147*, 260–270.

Hu, L. T., & Bentler, P. M. (1999). Cutoff criteria for fit indexes in covariance structure analysis: Conventional criteria versus new alternatives. *Structural Equation Modeling: A Multidisciplinary Journal, 6*(1), 1–55.

References

Hu, W., Shou, Y., Kang, M., & Park, Y. (2020). Risk management of manufacturing multinational corporations: The moderating effects of international asset dispersion and supply chain integration. *Supply Chain Management: An International Journal, 25*(1), 61–76.

Huang, M.-C., Yen, G.-F., & Liu, T.-C. (2014). Reexamining supply chain integration and the supplier's performance relationships under uncertainty. *Supply Chain Management: An International Journal, 19*(1), 64–78.

Hung, R. Y. Y., Yang, B., Lien, B.Y.-H., Mclean, G. N., & Kuo, Y.-M. (2010). Dynamic capability: Impact of process alignment and organizational learning culture on performance. *Journal of World Business, 45*(3), 285–294.

Huo, B., Zhao, X., & Lai, F. (2013). Supply chain quality integration: Antecedents and consequences. *IEEE Transactions on Engineering Management, 61*(1), 38–51.

Jia, F., & Rutherford, C. (2010). Mitigation of supply chain relational risk caused by cultural differences between China and the west. *International Journal of Logistics Management., 21*(2), 251–270.

Kang, M., Yang, M. G., Park, Y., & Huo, B. (2018). Supply chain integration and its impact on sustainability. *Industrial Management & Data Systems, 118*(9), 1749–1765.

Kauppi, K., Longoni, A., Caniato, F., & Kuula, M. (2016). Managing country disruption risks and improving operational performance: Risk management along integrated supply chains. *International Journal of Production Economics, 182*, 484–495.

Kull, T., Wiengarten, F., Power, D., & Shah, P. (2019). Acting as expected: Global leadership preferences and the pursuit of an integrated supply chain. *Journal of Supply Chain Management, 55*(3), 24–44.

Lai, F., Zhang, M., Lee, D. M. S., & Zhao, X. (2012). The impact of supply chain integration on mass customization capability: An extended resource-based view. *IEEE Transactions on Engineering Management, 59*(3), 443–456.

Lee, J.-H., & Kim, C.-O. (2008). Multi-agent systems applications in manufacturing systems and supply chain management: A review paper. *International Journal of Production Research, 46*(1), 233–265.

Lin, C. Y., & Ho, Y. H. (2009). RFID technology adoption and supply chain performance: An empirical study in China's logistics industry. *Supply Chain Management: An International Journal, 14*(5), 369–378.

Mullen, M. R. (1995). Diagnosing measurement equivalence in cross-national research. *Journal of International Business Studies, 26*(3), 573–596.

Narasimhan, R., & Kim, S. W. (2002). Effect of supply chain integration on the relationship between diversification and performance: Evidence from Japanese and Korean firms. *Journal of Operations Management, 20*(3), 303–323.

Park, Y., & Hong, P. (2017). Reshoring strategy: Case illustrations of Japanese manufacturing firms. In A. Vecchi (Ed.), *Reshoring of Manufacturing*. Measuring Operations Performance. Springer. https://doi.org/10.1007/978-3-319-58883-4_7

Park, Y., Ogawa, K., Tatsumoto, H., & Hong, P. (2009). The impact of product architecture on supply chain integration: A case study of Nokia and Texas Instruments. *International Journal of Services and Operations Management, 5*(6), 787–798.

Park, Y., Oh, J., & Fujimoto, T. (2012). Global expansion and supply chain integration: Case study of Korean firms. *International Journal of Procurement Management, 5*(4), 470–485.

Park, Y. W., & Hong, P. (2019). *Building network capabilities in turbulent competitive environments: Practices of global firms from Korea and Japan*. CRC Press.

Park, Y. W., Shintaku, J., & Hong, P. (2015). Effective supply chain integration: Case studies for Korean global firms in China. *International Journal of Manufacturing Technology and Management, 29*(3–4), 161–179.

Podsakoff, P. M., MacKenzie, S. B., Lee, J.-Y., & Podsakoff, N. P. (2003). Common method biases in behavioral research: A critical review of the literature and recommended remedies. *Journal of Applied Psychology, 88*(5), 879–903.

Rungtusanatham, M., Ng, C. H., Zhao, X., & Lee, T. S. (2008). Pooling data across transparently different groups of key informants: Measurement equivalence and survey research. *Decision Sciences, 39*(1), 115–145.

Shou, Y., Li, Y., Park, Y., & Kang, M. (2017). The impact of product complexity and variety on supply chain integration. *International Journal of Physical Distribution & Logistics Management, 47*(4), 297–317.

Shou, Y., Li, Y., Park, Y., & Kang, M. (2018). Supply chain integration and operational performance: The contingency effects of production systems. *Journal of Purchasing and Supply Management, 24*(4), 352–360.

Srinivasan, R., & Swink, M. (2015). Leveraging supply chain integration through planning comprehensiveness: An organizational information processing theory perspective. *Decision Sciences, 46*(5), 823–861.

Tarifa-Fernandez, J., & De Burgos-Jiménez, J. (2017). Supply chain integration and performance relationship: A moderating effects review. *International Journal of Logistics Management, 28*(4), 1243–1271.

Tian, M., Huo, B., Park, Y., & Kang, M. (2021). Enablers of supply chain integration: A technology-organization-environment view. *Industrial Management & Data Systems, 121*(8), 1871–1895.

Wagner, S. M., & Kemmerling, R. (2010). Handling nonresponse in logistics research. *Journal of Business Logistics, 31*(2), 357–381.

Williams, L. J., Hartman, N., & Cavazotte, F. (2010). Method variance and marker variables: A review and comprehensive CFA marker technique. *Organizational Research Methods, 13*(3), 477–514.

Zhao, X., Huo, B., Flynn, B. B., & Yeung, J. H. Y. (2008). The impact of power and relationship commitment on the integration between manufacturers and customers in a supply chain. *Journal of Operations Management, 26*(3), 368–388.

Zhao, X., Huo, B., Selen, W., & Yeung, J. H. Y. (2011). The impact of internal integration and relationship commitment on external integration. *Journal of Operations Management, 29*(1), 17–32.

Chapter 2
A Systematic Literature Review of Supply Chain Integration

Abstract Supply chain integration (SCI) has received considerable attention from the industry and academia in the last two decades. In this study, we aim to conduct a systematic literature review (SLR) and synthesize the extant literature on SCI. A sample of 187 research articles was identified through an SLR process. Descriptive analysis results were reported, including the time trend, journal outlets and research methods used in prior SCI studies. Moreover, we conduct a thematic analysis of the sampled articles to summarize the underpinning theories used in SCI research, key dimensions of SCI and performance improvements brought by SCI. Finally, we offer recommendations for future research in this area.

Keywords Supply chain integration · Systematic literature review · Thematic analysis

2.1 Introduction

As a core part of supply chain management, supply chain integration (SCI) has been a hot topic that scholars are keen to investigate (Flynn et al., 2010; Narayanan et al., 2011; Vickery et al., 2013). SCI manifests in terms of external integration and internal integration (Flynn et al., 2010). Integrating upstream suppliers and downstream customers as a way of external integration has been examined in prior studies (Ataseven & Nair, 2017). Meanwhile, internal integration is widely recognized as the enabler of external integration (Shou et al., 2017).

Discussion on the relationship between different dimensions of SCI and various performance has been ongoing since the mid-1990s (Huo et al., 2016; Jayaram et al., 2010; Schoenherr & Swink, 2012). Due to the outstanding work of SCI in firm performance improvements, SCI has been regarded as a key competitive differentiator that contributes to the firm's competitive advantages (Ataseven & Nair, 2017). This point of view has also been verified in practical applications. By adopting SCI, firms are enhanced to reduce production costs, improve product quality, and shorten cycle

---

This chapter is co-authored by Ziwei Yang and Yongyi Shou.

© The Author(s), under exclusive license to Springer Nature Singapore Pte Ltd. 2022
Y. Shou et al., *Supply Chain Integration for Sustainable Advantages*,
https://doi.org/10.1007/978-981-16-9332-8_2

time to improve production efficiency and elevate customer satisfaction (Flynn et al., 2010; Zhao et al., 2008).

Looking back at the research on SCI in the past 20 years, the fruitful results of prior studies have shown great vitality. Although scholars have conceptualized SCI from different perspectives, they generally agree that SCI is a strategic collaboration on intra- and inter-organizational activities with key supply chain partners (Childerhouse & Towill, 2011; Zhao et al., 2008). In recent years, due to the adoption of new technologies and the development of supply chain management research, SCI is no longer limited to upstream or downstream integration. Instead, it has gradually developed into an effective means to coordinate all supply chain partners to achieve real-time communication and integration through adopting digital supply chain platforms (Frank et al., 2019; Li et al., 2020). Not only the form of integration has been continuously broadened, but also the types of performance related to SCI. Initially, most SCI studies concentrate on operational or financial performance. Schoenherr and Swink (2012) demonstrated that there are direct relationships between SCI and financial performance and operational performance. Yet the performance effects of SCI are not conclusive. For example, Devaraj et al. (2007) found that supplier integration affects performance actively whereas customer integration does not have a significant impact on performance. There are also some aspects of performance that are less commonly addressed in the literature. For example, Liu et al. (2018) claimed that SCI benefits the adoption of green design strategy and hence helps realize better environmental performance. Besides performance effects, scholars also investigate the influence of SCI in other contexts. For example, Hu et al. (2020) have demonstrated that SCI played a moderating role in risk management.

By examining the research articles and literature reviews on SCI, we found that many articles have emphasized the importance of clarifying the dimensions of SCI and the performance improvements brought by SCI (Alfalla-Luque et al., 2013; Irani et al., 2017; van der Vaart & van Donk, 2008; Zhu et al., 2018). In this study, we focus on reviewing the extant literature about the dimensions of SCI and multiple aspects of performance to synthesize the existing findings. In brief, this study aims at answering the following research questions:

RQ1: What are the prevailing theories in the extant SCI research?
RQ2: Which dimensions of SCI are widely recognized in prior studies?
RQ3: What outcomes can SCI bring?

This chapter is structured as follows. Section 2.2 describes the review method used in this study. Then, the findings of descriptive analysis are provided in Sect. 2.3. Next, in Sect. 2.4 we present the results of thematic analysis and answer the three research questions raised. Finally, conclusions are provided in Sect. 2.5.

2.2 Method

In order to select, analyze and evaluate the literature related to specific research questions, a systematic literature review (SLR) was performed as SLR aids in knowledge building and development (Durach & Wiengarten, 2017; Tranfield et al., 2003). The SLR process is outlined in Table 2.1 and elaborated.

Stage 1—Identifying papers: The review was performed using the database of Web of Science Core Collection. Web of Science was one of the mainstream databases of peer-reviewed journals (Shashi et al., 2018). We selected "supply chain integration" as the topic keyword to identify the relevant literature and limited the time horizon from 2000 to 2020. Then we excluded review articles, resulting in a total of 1160 papers.

Stage 2—Paper selection and evaluation: Durach and Wiengarten (2017) emphasized the significance of article selection and evaluation criteria on review quality, so we used the 2021 edition of the Academic Journal Guide (AJG) and selected journals with AJG level 2 and above. We included the articles published in those journals in the fields of "Operations and Technology" and "Operations Research and Management Science". This process narrowed the sample to 351 articles. Then the abstracts of all 351 papers were screened. Studies not on SCI dimensions or outcomes were excluded. The sample was further reduced to 187 papers.

Table 2.1 The systematic literature review process

	Criteria	Detailed steps
Stage 1: Identifying papers	1.1 Keywords search in *Web of Science Core Collection* (1218 papers)	Search relevant articles in *Web of Science Core Collection* with keyword "supply chain integration", excluding "Review articles"
	1.2 Limit the time horizon from 2000 to 2020 (1160 papers)	Select papers published between 2000 and 2020
Stage 2: Paper selection and evaluation	2.1 Included papers from AJG 2021 level 2 and above journals (351 papers)	Select level 2 and above journals from AJG 2021 in the fields "Operations and Technology" and "Operations Research and Management Science"
	2.2 Excluded irrelevant articles after abstract analysis (187 papers)	Papers not on SCI dimensions or SCI outcomes were excluded
Stage 3: Paper synthesis and results reporting	3.1 Descriptive analysis	Graphs and tables were designed to report information of selected publications
	3.2 Thematic analysis	Thematic reporting of articles to answer the research questions

Stage 3—Paper synthesis and results reporting: A data extraction form which recorded the contents from the 187 papers was used to aid the analysis of the papers. A descriptive analysis was provided to show the following information: year, journal, and methodology. We want to present the popularity of this topic (i.e., SCI), the trend of publication volume, and the journals that favor this topic, to provide some references for researchers in need. The results of descriptive analysis were provided in the following section. Furthermore, a thematic analysis reports the findings in detail and help in drawing conclusions. The thematic analysis, as mentioned above, discuss theories used in SCI research, the dimensions of SCI, and SCI outcomes.

2.3 Descriptive Analysis

2.3.1 Publication Distributions

We display the distribution of papers by journal and year in this section. Figure 2.1 shows the publication year of the 187 articles in our sample. It can be seen that there is a steady increase of publications since 2000. Although there was a slight decline in 2017, the number of articles has maintained a relatively high level since 2018. This rebound may be due to the emerging new technologies and the corresponding expansion of supply chain management studies, which extend SCI research into new contexts. For instance, SCI has become an effective way to improve environmental performance in green supply chains (Wong et al., 2020, 2021). SCI is also an efficient means to mitigate supply chain risks (Chaudhuri et al., 2018; Hu et al., 2020; Jajja et al., 2018; Munir et al., 2020). With the help of information and communication technology (ICT), SCI can help realize real-time communication and information exchange between key partners in the whole supply chain (Vanpoucke et al., 2017; Wei et al., 2020). Therefore, SCI has received much attention in recent years.

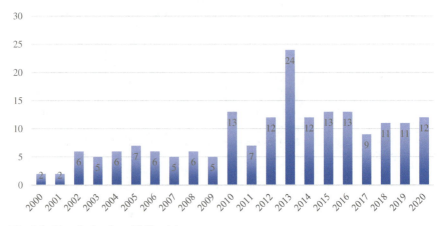

Fig. 2.1 Historical series of SCI articles

2.3 Descriptive Analysis

Fig. 2.2 Journals publishing SCI studies

Figure 2.2 reports the journals that published SCI studies from 2000 to 2020. The top contributor is *International Journal of Production Economics* (35 papers), followed by *Supply Chain Management: An International Journal* (31 papers) and *International Journal of Operations & Production Management* (23 papers). Besides, *Journal of Operations Management, Production and Operations Management*, and *European Journal of Operational Research*, recognized as top journals by peers, also published a number of SCI-related articles.

2.3.2 Research Methods

Table 2.2 lists the research methods of SCI studies and examples of corresponding research using those methods. We selected 45 mostly cited articles and 45 recently published articles and obtained 74 articles after removing duplicates between the two collections. Following Jia et al. (2021), we divided the research methods into four categories: theoretical, analytical, empirical studies and meta-analysis.

Theoretical: No empirical data is provided in this category of articles. They contain only theoretical work like finding evidence from past research to support an idea. For example, Harland et al. (2007) showed the potential benefits of SCI and the contribution of e-Business in information integration by reviewing previous literature. Some scholars just propose a framework or provide strategies, such as Nandi et al. (2020) who developed a framework to help firms integrate blockchain technology

Table 2.2 Research methods in SCI research

Research method	Number of articles	Typical studies
Theoretical		
Conceptual	6	Harland et al. (2007), Mikkola and Skjott-Larsen (2004), Nandi et al. (2020)
Analytical		
Modeling	3	Krajewski and Wei (2001), Kull et al. (2018)
Empirical		
Survey	55	Tarifa-Fernandez et al. (2019), Turkulainen et al. (2017)
Case study	2	Palomero and Chalmeta (2014), Silvestro and Lustrato (2014)
Mixed method	5	Bernon et al. (2013), Jin et al. (2013), Richey et al. (2009)
Meta-analysis	2	Ataseven and Nair (2017)

into their supply chain according to the resource-based view. Mikkola and Skjott-Larsen (2004) introduced three strategies for SCI management: mass customization, postponement, and modularization.

Analytical: All papers in this category use a method of logical thinking to explore research models, usually represented by functions or theorems. Although this kind of method is not commonly used, Zhao et al. (2011) adopted mixed integer programming (MIP) and simulation method to prove that SCI can reduce costs and improve customer service performance. Krajewski and Wei (2001) explored the impact of different levels of SCI on costs through a stochastic model.

Empirical: This is the largest category of papers in our sample. All empirical papers from the sample collect data through survey, case study, or mixed methods. Survey is the most popular method because of its ability to quantify some features of SCI. For instance, Tarifa-Fernandez et al. (2019) used hierarchical multiple regression to analyze the relationship between absorptive capacity, external SCI and supply chain performance by analyzing 99 responses from the horticulture marketing department. In addition to survey, case study is of great help to understand SCI deeply and comprehensively. Silvestro and Lustrato (2014) did a case study of two international banks and reported their role in improving SCI. In another case study, Palomero and Chalmeta (2014) applied multi-case study method to 30 industrial small and medium-sized enterprises (SMEs), which aims at guiding SMEs to overcome the main barriers in achieving SCI. Some scholars use multiple methods. For example, Bernon et al. (2013) analyzed a case composed of an original equipment manufacturer (OEM) and two retailers and use semi-structured interview method and observations to demonstrate the benefits and obstacles of SCI in the context of the retail product return process. Chang et al. (2013) used interview and questionnaire survey to illustrate the role of e-procurement in improving supply chain performance.

2.3 Descriptive Analysis   15

Meta-analysis: It is a specific type of empirical study that obtains secondary data from prior studies. Ataseven and Nair (2017) used meta-analysis to investigate the relationships between SCI and various performance dimensions. Their results show that internal integration, supplier integration and customer integration have positive impact on the company's financial performance.

2.4 Thematic Analysis

We did a text analysis of the sampled articles by Leximancer, which is an automated text mining analysis software (Angus et al., 2013). Leximancer regards concepts as a collection of words which always appear together. Leximancer clusters similar concepts into topics, represented by colored circles in the concept map (Montecchi et al., 2021). The overlap between the circles indicates that these topics are usually discussed together. The line between two concepts implies that there is a connection between them. We used Leximancer to perform a text analysis on the abstracts of the sampled articles. The result is presented in Fig. 2.3.

The thematic analysis reveals that internal integration, supplier integration and customer integration are essential components of SCI. The result of text analysis confirms that these articles have investigated the dimensions of SCI and the relationships between SCI and firm performance.

Next, we aim to answer the three research questions raised in this review. First, we will explore the theories applied in extant SCI research, then show the dimensions of SCI and the performance that can be brought by SCI.

2.4.1 Theories in SCI Research

Among the 187 articles that were included in the final sample, 38 of them have theories that were explicitly stated in the abstract. Table 2.3 shows the 22 theories mentioned in these articles. Resource-based view (RBV), organizational information processing theory (OIPT), contingency theory, and social capital theory (SCT) are the top four theories. It is noted that a few papers employed multiple theories in their study. Next, we will briefly introduce the four theories and their applications in SCI research.

Resource-based view: The resource-based view of the firm argues that a resource must be valuable, inimitable, and scarce for it to transform into competitive advantages (Barney, 1991). Based on the RBV, Huo et al. (2016) showed the role of human capital in SCI and its connection with competitive performance. RBV not only regard productive resources as competitive advantages, but also concerned about invisible resources that do not work directly. For example, Mitra and Singhal (2008) regarded the membership of an industry exchange that is not open to the public as a competitive advantage, believing that the membership can help reduce transaction costs and

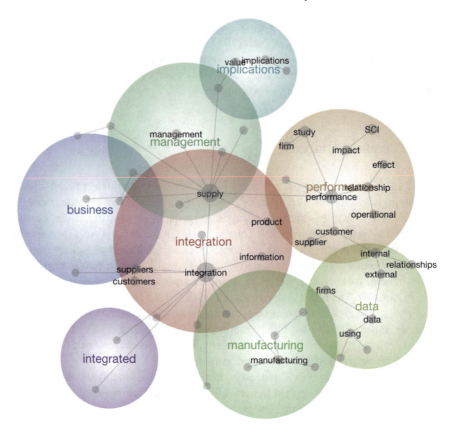

Fig. 2.3 Concept map

promote cooperation. According to RBV, Shee et al. (2018) believed that ICT can be seen as a technological resource, which can be combined with other organizational capabilities to drive the company to achieve its performance goals.

In addition, Lai et al. (2012) applied the extended resource-based view (ERBV) to explore the effect of SCI on mass customization capability. ERBV suggests that firms' critical resources not only accumulate within but also extend beyond their boundaries (Lavie, 2006). Moreover, it indicates that a firm's internal capabilities such as SCI can help facilitate the exploitation of external resources, and thus could further enhance firm performance (Lewis et al., 2010). Recently, ERBV has been increasingly adopted by supply chain management research (Shou et al., 2021; Yang et al., 2019).

Organizational information processing theory: OIPT indicates that an organization can improve performance by improving the fit between its information processing capabilities and information processing requirements (Premkumar et al., 2005; Tushman & Nadler, 1978). Srinivasan and Swink (2015) took OIPT as a theoretical perspective to argue that SCI enables information-intensive exchange between

2.4 Thematic Analysis

Table 2.3 Theories used in SCI research

Theory	Number of papers	Studies
Resource-based view	10	Arellano et al. (2019), Cousins and Menguc (2006), Leuschner et al. (2013), Davis et al. (2014), Mitra and Singhal (2008), Mora-Monge et al. (2019), Nandi et al. (2020), Schoenherr and Swink (2012), Shee et al. (2018), Tarifa-Fernandez et al. (2019)
Organizational information processing theory	8	Gu et al. (2017), Li et al. (2020), Schoenherr and Swink (2012), Shou, Hu, et al. (2018), Shou, Li, et al. (2018), Srinivasan and Swink (2015), Terjesen et al. (2012), Wong et al. (2011)
Contingency theory	4	Ebrahimi et al. (2018), Davis et al. (2014); Terjesen et al. (2012), Wong et al. (2011)
Social capital theory	3	Horn et al. (2014), Jacobs et al. (2016), Mora-Monge et al. (2019)
Coordination theory	2	Ebrahimi et al. (2018), Terjesen et al. (2012)
Dynamic capabilities view	2	Mora-Monge et al. (2019)
Organizational capability theory	2	Huo (2012), Qi et al. (2017)
Relational view	2	Leuschner et al. (2013)
Resource dependence theory	2	Lii and Kuo (2016), Shah et al. (2020)
Trust theory	1	He et al. (2014)
Adaptive structuration theory	1	Droge et al. (2012)
Ambidexterity theory	1	Wong et al. (2013)
Behavioral theory of the firm	1	Durach and Wiengarten (2020)
Extended resource-based view	1	Lai et al. (2012)
Institutional theory	1	Turkulainen et al. (2017)
Knowledge-based view	1	Gu et al. (2017)
Network theory	1	Stolze et al. (2018)
Organizational learning theory	1	Yu et al. (2013)
Resource orchestration theory	1	Liu et al. (2016)
Service-dominant logic	1	Jitpaiboon et al. (2013)
Strategic leadership theory	1	Birasnav and Bienstock (2019)
Transaction cost theory	1	Huang et al. (2014)

supply chain partners. Organizations need more information when performing tasks and SCI facilitate providing more accurate information. Based on OIPT, Shou, Hu, et al. (Shou, Hu, et al., 2018) confirmed that the required quality, cost and other performance can be achieved through external integration only with proper production systems. Li et al. (2020) discussed the impact of digital supply chain platforms on economic and environmental performance through the lens of OIPT, which helps to understand the changes brought by new information technologies.

Contingency theory: Contingency theory believes that to achieve the goal of maximizing performance, an organization should ensure that its strategy, structure, and process match its environment (Miller, 1992; Sinha & Van de Ven, 2005). Uncertainty in technology and demand will regulate internal integration and customer delivery performance. Boon-itt and Wong (2011) confirmed the positive effects of supplier integration and internal integration on customer delivery performance as SCI can enhances the firm's environmental adaptability. Ebrahimi et al. (2018) examined the mediating effects of SCI on the relationships in organizational structure from a contingency perspective.

Social capital theory: SCT believes that social structure can enable social actions to generate or exchange assets and create collectively owned social assets. It is also the "relationship glue" between social actors, such as employees, suppliers, and buyers (Koka & Prescott, 2002; Nahapiet & Ghoshal, 1998). From the perspective of SCT, Horn et al. (2014) explored the relationship between cognitive capital, structural capital and relationship capital, internal and external integration, and the success of global procurement. Based on SCT, Mora-Monge et al. (2019) argued that SCI can improve the trust of trading partners, thus promote communication, knowledge transfer and risk mitigation in the supply chain network.

2.4.2 Dimensions of SCI

Following prior studies, we divide the SCI into three dimensions: internal integration, supplier integration and customer integration (Flynn et al., 2010; Zhao et al., 2011).

Internal integration refers to the extent to which an organization has structured the procedures, practices, and behaviors of its internal functional units to achieve mutual collaboration and synchronization for fulfilling customer requirements (Horn et al., 2014; Huo, 2012). Therefore, managers usually adopt methods such as integrating inventory management, making a sales forecast with multiple departments, and establishing cross-functional teams when conducting internal integration (Munir et al., 2020; Vickery et al., 2003; Wong et al., 2021). The communication by information sharing and cooperation by joint decision-making among the functional units or departments are two key features of internal integration. The joint decision-making and risk management realized by information sharing within a company have a huge effect on the company's performance.

2.4 Thematic Analysis

External integration can be further divided into supplier integration and customer integration (Droge et al., 2004; Koufteros et al., 2007), which refers to a partnership with its key customers and suppliers. Supplier integration includes information sharing, joint planning, and partnerships establishment with suppliers (Zhang et al., 2018). Supplier integration attaches great importance to the role of information exchange. Based on information exchange, integrated supply chains are able to conduct quick ordering, vendor managed inventory (VMI), just-in-time (JIT) purchasing, stock level control, risk pooling and revenue sharing (Droge et al., 2012; Narasimhan & Kim, 2002; Schoenherr & Swink, 2012). Establishing strategic relationships between suppliers can promote SCI. A close supplier relationship can be achieved through making joint formulation of strategies and joint decisions (Kim, 2009; Prajogo et al., 2012).

Customer integration needs to conduct close information exchange and share information with key customers. Customer integration can help to better understand customer needs and adjust organizational functions (Koufteros et al., 2007). Collaboration with key customers in product design, process management and joint decision-making can reduce uncertainties about customer expectations. In product and process development, companies use a variety of methods to understand their customers. Customers may get involved in activities ranging from idea generation to production planning, order tracking and tracing, and delivery of products (Munir et al., 2020; Wiengarten et al., 2019).

To conduct empirical studies on SCI, scholars have developed scales to measure the individual dimensions of SCI in questionnaire survey. Table 2.4 summarizes the typical scales that are used in the extant literature to measure the three key dimensions of SCI.

International Manufacturing Strategy Survey (IMSS) is one of the well-known questionnaire surveys in SCI research. The questionnaire has been updated every four to five years since 1992 (Wiengarten et al., 2014) and now there are six rounds of IMSS. A group of researchers designed the scales in IMSS based on existing literature, which guarantees that the scales are reliable and representative. Wiengarten et al. (2019) have compared the scales of SCI in the six rounds of IMSS and found that the scales are all similar from the third to the fifth rounds. In the sixth round IMSS, the scales had been adjusted according to the latest SCI literature, as presented in Table 2.4.

2.4.3 SCI and Performance

Supply chain integrative capabilities are drivers of firm performance (Huo, 2012). The extant literature on SCI has reported the effects of SCI on a number of performance dimensions, mainly in financial, operational and environmental performance. The performance improvements that can be brought by SCI are listed in Table 2.5.

Table 2.4 Dimensions of SCI

Dimension	Study	Scale
Internal integration	Narasimhan and Kim (2002)	1. Data integration among internal functions through information network 2. System-wide information system integration among internal functions 3. Real-time searching of the level of inventory 4. Real-time searching of logistics-related operating data 5. Data integration in production process 6. Integrative inventory management 7. The construction of system-wide interaction system between production and sales 8. The utilization of periodic interdepartmental meetings among internal function
	Wong et al. (2020)	1. Integrated environmental and business strategy 2. Internal integrated environmental management system 3. Cross-functional collaboration for environmental management
	IMSS	1. Sharing information with purchasing department 2. Joint decision making with purchasing department 3. Sharing information with sales department 4. Joint decision making with sales department
Supplier integration	Wong et al. (2020)	1. Exchange environmental information with suppliers 2. Provide environmental assistance to suppliers 3. Integrate environmental management process with suppliers 4. Environmental collaboration with suppliers
	IMSS	1. Sharing information with key suppliers 2. Developing collaborative approaches with key suppliers 3. Joint decision-making with key suppliers/customers 4. System coupling with key suppliers
Customer integration	Wong et al. (2020)	1. Exchange environmental information with customers 2. Integrate environmental management process with customers 3. Environmental collaboration with customers
	IMSS	1. Sharing information with key customers 2. Developing collaborative approaches with key customers 3. Joint decision-making with key customers 4. System coupling with key suppliers/customers

2.4 Thematic Analysis

Table 2.5 Performance effects of SCI

Study	Sample	Data source	SCI dimensions	Performance dimensions	Direct effect
Boon-itt and Wong (2011)	734 firms	Thailand	Internal, supplier, and customer integration	Operational performance and financial performance	Positive
Chen et al. (2013)	117 supply chain executives	America	Internal, supplier, and customer integration	Agility performance	Positive
Droge et al. (2012)	57 suppliers	North America	Supplier, and customer integration	Operational performance	Positive
Jacobs et al. (2016)	214 manufacturers	China	Internal, supplier, and customer integration	Employee satisfaction	Positive
Kim (2009)	1490 questionnaire results	Korea and Japan	Internal, supplier, and customer integration	Financial performance and operational performance	Positive
Munir et al. (2020)	931 samples from IMSS	Multi-country	Internal, supplier, and customer integration	Operational performance	Positive
Narasimhan and Kim (2002)	623 manufacturing organizations	Korea and Japan	Internal, supplier, and customer integration	Financial performance	Positive
Prajogo and Olhager (2012)	232 firms	Australia	Internal, supplier, and customer integration	Financial performance and operational performance	Positive
Schoenherr and Swink (2012)	403 samples	Multi-country	Internal, supplier, and customer integration	Financial performance and operational performance	Positive
Sezen (2008)	125 firms	Turkey	Internal, supplier, and customer integration	Operational performance and financial performance	Non-significant

(continued)

Table 2.5 (continued)

Study	Sample	Data source	SCI dimensions	Performance dimensions	Direct effect
Vickery et al. (2003)	57 suppliers	North America	Internal, supplier, and customer integration	Financial performance and operational performance	Positive
Wiengarten et al. (2019)	293 samples from IMSS	Multi-country	Supplier and customer integration	Financial performance and operational performance	Positive
Wong et al. (2020)	1000 firms	China	Green internal, supplier, and customer integration	Environmental performance, innovation performance and financial performance	Positive

The literature has found the role of SCI in achieving financial performance (Flynn et al., 2010; Kim, 2009; Wong et al., 2020). Internal integration can improve financial performance. Swink et al. (2007) found that strategy integration and product-process integration influence financial performance. Similarly, external integration can influence financial performance. For example, Narasimhan and Kim (2002) found that supplier integration contributes to financial performance. However, the impact of customer integration on financial performance is inconsistent, while some studies failed to find a significant relationship between customer integration and financial performance (Flynn et al., 2010).

Operational performance covers several aspects such as quality, flexibility, cost reduction and customer service (Chavez et al., 2015). Positive impact of SCI on operational performance has been widely observed (Devaraj et al., 2007). Both internal and external integration can improve process efficiency, product effectiveness and logistics service performance (Germain et al., 2008; Saeed et al., 2005). Chen et al. (2013) confirmed that integrating the supply chain can improve supply chain agility and reduce the risk of the supply chain at the same time. In general, SCI is an excellent method to deal with complex supply chains. It can realize efficient information exchange between supply chains, increasing the satisfaction of supply chain partners, reducing the overall risk of the supply chain, and improving production efficiency as well as transportation efficiency.

In addition to the most frequently discussed operational performance and financial performance, environmental performance is closely related to SCI. In green supply chain management (GSCM), Wong et al. (2020) has found that the three dimensions of green SCI (i.e., internal, supplier and customer integration) can improve cost performance and environmental performance. Customer integration

2.5 Conclusions

We sample and summarize 187 papers through an SLR process. By performing descriptive analysis, we illustrate the trend of SCI publications. By introducing well-known theories in SCI research, we aim to understand the viewpoints on SCI in the academia. By reporting the dimensions of SCI and performance improvements brought by SCI, we show the practical significance of SCI in the industry.

Since the theories underpinning SCI are still developing, this study attempts to give a brief overview of the popular theories used in SCI. As an effective approach to improve supply chain capabilities, SCI has been regarded as a source of competitive advantages, and therefore RBV is the most widely used theory in SCI research. While RBV can help explain why SCI improves performance, OIPT explains how internal integration contributes to external integration and increasing the latter's effectiveness (Schoenherr & Swink, 2012). OIPT pinpoints the importance to align the firm's information processing capability with the information processing needs, which are largely determined by their supply chain partners. Both theories provide reliable theoretical basis for explaining the impact of SCI on firm performance.

Moreover, we provide insights into the relationships between the dimensions of SCI and its performance outcomes. No matter internal integration, supplier integration or customer integration, each dimension is essential for supply chain management. Internal integration contributes to coordinating the company and the supply chain partners, while external integration emphasizes the establishment of close cooperation with suppliers and customers through information exchange and common practices (Ataseven & Nair, 2017). SCI promotes a smoother and more reliable material flow and information flow in the supply chain. SCI also facilitates the implementation of management practices including quality management (Shou et al., 2020). Both operational and financial performance can be enhanced through the implementation of SCI, which is in line with previous literature reviews (Alfalla-Luque et al., 2013; Danese et al., 2013; Irani et al., 2017). It is worth mentioning that SCI can improve the environmental performance of green supply chains (Wong et al., 2020). GSCM has been a very important issue after the Paris Agreement in 2015, so we expect more research on green SCI.

In this literature review, we chose the most cited and recently published articles for detailed analysis, which discloses the widely recognized ideas and the latest research on SCI. Because of emerging new technologies and the corresponding expansion of supply chain management research, the performance brought by SCI and the implementation methods of SCI have become more diversified. SCI undoubtedly affects the relationships between the key partners in the supply chain. However, there is a dearth of research on SCI's impact on social performance in the extant literature. It is a great opportunity to make up the shortfall of studies on this impact in the future. Finally, as revealed by this SLR, previous empirical studies mainly utilized data obtained from questionnaire surveys. Other data sources may offer new opportunities on SCI research, particularly in those emerging areas.

References

Alfalla-Luque, R., Medina-Lopez, C., & Dey, P. K. (2013). Supply chain integration framework using literature review. *Production Planning & Control, 24*(8–9), 800–817.

Angus, D., Rintel, S., & Wiles, J. (2013). Making sense of big text: A visual-first approach for analysing text data using Leximancer and Discursis. *International Journal of Social Research Methodology, 16*(3), 261–267.

Arellano, M. C., Rebolledo, C., & Tao, Z. (2019). Improving operational plant performance in international manufacturing networks: The effects of integrative capabilities and plant roles. *Production Planning & Control, 30*(2–3), 112–130.

Ataseven, C., & Nair, A. (2017). Assessment of supply chain integration and performance relationships: A meta-analytic investigation of the literature. *International Journal of Production Economics, 185*, 252–265.

Barney, J. B. (1991). Firm resources and sustained competitive advantage. *Journal of Management, 17*(1), 99–120.

Bernon, M., Upperton, J., Bastl, M., & Cullen, J. (2013). An exploration of supply chain integration in the retail product returns process. *International Journal of Physical Distribution & Logistics Management, 43*(7), 586–608.

Birasnav, M., & Bienstock, J. (2019). Supply chain integration, advanced manufacturing technology, and strategic leadership: An empirical study. *Computers & Industrial Engineering, 130*, 142–157.

Boon-itt, S., & Wong, C. Y. (2011). The moderating effects of technological and demand uncertainties on the relationship between supply chain integration and customer delivery performance. *International Journal of Physical Distribution & Logistics Management, 41*(3), 253–276.

Chang, H., Tsai, Y.-C., & Hsu, C.-H. (2013). E-procurement and supply chain performance. *Supply Chain Management: An International Journal, 18*(1), 34–51.

Chaudhuri, A., Boer, H., & Taran, Y. (2018). Supply chain integration, risk management and manufacturing flexibility. *International Journal of Operations & Production Management, 38*(3), 690–712.

Chavez, R., Yu, W., Gimenez, C., Fynes, B., & Wiengarten, F. (2015). Customer integration and operational performance: The mediating role of information quality. *Decision Support Systems, 80*, 83–95.

Chen, D. Q., Preston, D. S., & Xia, W. (2013). Enhancing hospital supply chain performance: A relational view and empirical test. *Journal of Operations Management, 31*(6), 391–408.

Childerhouse, P., & Towill, R. (2011). Arcs of supply chain integration. *International Journal of Production Research, 49*(24), 7441–7468.

Cousins, P. D., & Menguc, B. (2006). The implications of socialization and integration in supply chain management. *Journal of Operations Management, 24*(5), 604–620.

Danese, P., Romano, P., & Formentini, M. (2013). The impact of supply chain integration on responsiveness: The moderating effect of using an international supplier network. *Transportation Research Part E: Logistics and Transportation Review, 49*(1), 125–140.

Davis, J. M., Mora-Monge, C., Quesada, G., & Gonzalez, M. (2014). Cross-cultural influences on e-value creation in supply chains. *Supply Chain Management: An International Journal, 19*(2), 187–199.

References 25

Devaraj, S., Krajewski, L., & Wei, J. C. (2007). Impact of eBusiness technologies on operational performance: The role of production information integration in the supply chain. *Journal of Operations Management, 25*(6), 1199–1216.

Droge, C., Jayaram, J., & Vickery, S. K. (2004). The effects of internal versus external integration practices on time-based performance and overall firm performance. *Journal of Operations Management, 22*(6), 557–573.

Droge, C., Vickery, S. K., & Jacobs, M. A. (2012). Does supply chain integration mediate the relationships between product/process strategy and service performance? An empirical study. *International Journal of Production Economics, 137*(2), 250–262.

Durach, C. F., & Wiengarten, F. (2017). Exploring the impact of geographical traits on the occurrence of supply chain failures. *Supply Chain Management: An International Journal, 22*(2), 160–171.

Durach, C. F., & Wiengarten, F. (2020). Supply chain integration and national collectivism. *International Journal of Production Economics, 224*, 107543.

Ebrahimi, S. M., Koh, S. C. L., Genovese, A., & Kumar, N. (2018). Structure-integration relationships in oil and gas supply chains. *International Journal of Operations & Production Management, 38*(2), 424–445.

Flynn, B. B., Huo, B., & Zhao, X. (2010). The impact of supply chain integration on performance: A contingency and configuration approach. *Journal of Operations Management, 28*(1), 58–71.

Frank, A. G., Dalenogare, L. S., & Ayala, N. F. (2019). Industry 4.0 technologies: Implementation patterns in manufacturing companies. *International Journal of Production Economics, 210*, 15–26.

Germain, R., Claycomb, C., & Droge, C. (2008). Supply chain variability, organizational structure, and performance: The moderating effect of demand unpredictability. *Journal of Operations Management, 26*(5), 557–570.

Gu, Q., Jitpaipoon, T., & Yang, J. (2017). The impact of information integration on financial performance: A knowledge-based view. *International Journal of Production Economics, 191*, 221–232.

Harland, C. M., Caldwell, N. D., Powell, P., & Zheng, J. (2007). Barriers to supply chain information integration: SMEs adrift of eLands. *Journal of Operations Management, 25*(6), 1234–1254.

He, Y., Keung Lai, K., Sun, H., & Chen, Y. (2014). The impact of supplier integration on customer integration and new product performance: The mediating role of manufacturing flexibility under trust theory. *International Journal of Production Economics, 147*, 260–270.

Horn, P., Scheffler, P., & Schiele, H. (2014). Internal integration as a pre-condition for external integration in global sourcing: A social capital perspective. *International Journal of Production Economics, 153*, 54–65.

Hu, W. J., Shou, Y. Y., Kang, M. G., & Park, Y. (2020). Risk management of manufacturing multinational corporations: The moderating effects of international asset dispersion and supply chain integration. *Supply Chain Management: An International Journal, 25*(1), 61–76.

Huang, M.-C., Yen, G.-F., & Liu, T.-C. (2014). Reexamining supply chain integration and the supplier's performance relationships under uncertainty. *Supply Chain Management: An International Journal, 19*(1), 64–78.

Huo, B. (2012). The impact of supply chain integration on company performance: An organizational capability perspective. *Supply Chain Management: An International Journal, 17*(6), 596–610.

Huo, B., Ye, Y., Zhao, X., & Shou, Y. (2016). The impact of human capital on supply chain integration and competitive performance. *International Journal of Production Economics, 178*, 132–143.

Irani, Z., Kamal, M. M., Sharif, A., & Love, P. E. D. (2017). Enabling sustainable energy futures: Factors influencing green supply chain collaboration. *Production Planning & Control, 28*(6–8), 684–705.

Jacobs, M. A., Yu, W., & Chavez, R. (2016). The effect of internal communication and employee satisfaction on supply chain integration. *International Journal of Production Economics, 171*, 60–70.

Jajja, M. S. S., Chatha, K. A., & Farooq, S. (2018). Impact of supply chain risk on agility performance: Mediating role of supply chain integration. *International Journal of Production Economics, 205*, 118–138.

Jayaram, J., Tan, K.-C., & Nachiappan, S. P. (2010). Examining the interrelationships between supply chain integration scope and supply chain management efforts. *International Journal of Production Research, 48*(22), 6837–6857.

Jia, M., Stevenson, M., & Hendry, L. (2021). A systematic literature review on sustainability-oriented supplier development. *Production Planning & Control*, 1–21. https://doi.org/10.1080/09537287.2021.1958388

Jin, Y. H., Fawcett, A. M., & Fawcett, S. E. (2013). Awareness is not enough commitment and performance implications of supply chain integration. *International Journal of Physical Distribution & Logistics Management, 43*(3), 205–230.

Jitpaiboon, T., Dobrzykowski, D. D., Ragu-Nathan, T. S., & Vonderembse, M. A. (2013). Unpacking IT use and integration for mass customisation: A service-dominant logic view. *International Journal of Production Research, 51*(8), 2527–2547.

Kim, S. W. (2009). An investigation on the direct and indirect effect of supply chain integration on firm performance. *International Journal of Production Economics, 119*(2), 328–346.

Koka, B. R., & Prescott, J. E. (2002). Strategic alliances as social capital: A multidimensional view. *Strategic Management Journal, 23*(9), 795–816.

Koufteros, X. A., Cheng, T. C. E., & Lai, K.-H. (2007). "Black-box" and "gray-box" supplier integration in product development: Antecedents, consequences and the moderating role of firm size. *Journal of Operations Management, 25*(4), 847–870.

Krajewski, L., & Wei, J. C. (2001). The value of production schedule integration in supply chains. *Decision Sciences, 32*(4), 601–634.

Kull, T. J., Kotlar, J., & Spring, M. (2018). Small and medium enterprise research in supply chain management: The case for single-respondent research designs. *Journal of Supply Chain Management, 54*(1), 23–34.

Lai, F., Zhang, M., Lee, D. M. S., & Zhao, X. (2012). The impact of supply chain integration on mass customization capability: An extended resource-based view. *IEEE Transactions on Engineering Management, 59*(3), 443–456.

Lavie, D. (2006). The competitive advantage of interconnected firms: An extension of the resource-based view. *Academy of Management Review, 31*(3), 638–658.

Leuschner, R., Rogers, D. S., & Charvet, F. F. (2013). A meta-analysis of supply chain integration and firm performance. *Journal of Supply Chain Management, 49*(2), 34–57.

Lewis, M., Brandon-Jones, A., Slack, N., & Howard, M. (2010). Competing through operations and supply: The role of classic and extended resource-based advantage. *International Journal of Operations & Production Management, 30*(10), 1032–1058.

Li, Y., Dai, J., & Cui, L. (2020). The impact of digital technologies on economic and environmental performance in the context of Industry 4.0: A moderated mediation model. *International Journal of Production Economics, 229*, 107777.

Lii, P., & Kuo, F.-I. (2016). Innovation-oriented supply chain integration for combined competitiveness and firm performance. *International Journal of Production Economics, 174*, 142–155.

Liu, H., Wei, S., Ke, W., Wei, K. K., & Hua, Z. (2016). The configuration between supply chain integration and information technology competency: A resource orchestration perspective. *Journal of Operations Management, 44*, 13–29.

Liu, Y., Blome, C., Sanderson, J., & Paulraj, A. (2018). Supply chain integration capabilities, green design strategy and performance: A comparative study in the auto industry. *Supply Chain Management: An International Journal, 23*(5), 431–443.

Mikkola, J. H., & Skjott-Larsen, T. (2004). Supply-chain integration: Implications for mass customization, modularization and postponement strategies. *Production Planning & Control, 15*(4), 352–361.

Miller, D. (1992). Environmental fit versus internal fit. *Organization Science, 3*(2), 159–178.

Mitra, S., & Singhal, V. (2008). Supply chain integration and shareholder value: Evidence from consortium based industry exchanges. *Journal of Operations Management, 26*(1), 96–114.

Montecchi, M., Plangger, K., & West, D. C. (2021). Supply chain transparency: A bibliometric review and research agenda. *International Journal of Production Economics, 238*, 108152.

Mora-Monge, C., Quesada, G., Gonzalez, M. E., & Davis, J. M. (2019). Trust, power and supply chain integration in web-enabled supply chains. *Supply Chain Management: An International Journal, 24*(4), 524–539.

Munir, M., Jajja, M. S. S., Chatha, K. A., & Farooq, S. (2020). Supply chain risk management and operational performance: The enabling role of supply chain integration. *International Journal of Production Economics, 227*, 107667.

Nahapiet, J., & Ghoshal, S. (1998). Social capital, intellectual capital, and the organizational advantage. *Academy of Management Review, 23*(2), 242–266.

Nandi, M. L., Nandi, S., Moya, H., & Kaynak, H. (2020). Blockchain technology-enabled supply chain systems and supply chain performance: A resource-based view. *Supply Chain Management: An International Journal, 25*(6), 841–862.

Narasimhan, R., & Kim, S. W. (2002). Effect of supply chain integration on the relationship between diversification and performance: Evidence from Japanese and Korean firms. *Journal of Operations Management, 20*(3), 303–323.

Narayanan, S., Jayaraman, V., Luo, Y., & Swaminathan, J. M. (2011). The antecedents of process integration in business process outsourcing and its effect on firm performance. *Journal of Operations Management, 29*(1–2), 3–16.

Palomero, S., & Chalmeta, R. (2014). A guide for supply chain integration in SMEs. *Production Planning & Control, 25*(5), 372–400.

Prajogo, D., Huo, B. F., & Han, Z. J. (2012). The effects of different aspects of ISO 9000 implementation on key supply chain management practices and operational performance. *Supply Chain Management: An International Journal, 17*(3), 306–322.

Prajogo, D., & Olhager, J. (2012). Supply chain integration and performance: The effects of long-term relationships, information technology and sharing, and logistics integration. *International Journal of Production Economics, 135*(1), 514–522.

Premkumar, G., Ramamurthy, K., & Saunders, C. S. (2005). Information processing view of organizations: An exploratory examination of fit in the context of interorganizational relationships. *Journal of Management Information Systems, 22*(1), 257–294.

Qi, Y., Huo, B., Wang, Z., & Yeung, H. Y. J. (2017). The impact of operations and supply chain strategies on integration and performance. *International Journal of Production Economics, 185*, 162–174.

Richey, R. G., Chen, H. Z., Upreti, R., Fawcett, S. E., & Adams, F. G. (2009). The moderating role of barriers on the relationship between drivers to supply chain integration and firm performance. *International Journal of Physical Distribution & Logistics Management, 39*(9–10), 826–840.

Saeed, K. A., Malhotra, M. K., & Grover, V. (2005). Examining the impact of interorganizational systems on process efficiency and sourcing leverage in buyer-supplier dyads. *Decision Sciences, 36*(3), 365–396.

Schoenherr, T., & Swink, M. (2012). Revisiting the arcs of integration: Cross-validations and extensions. *Journal of Operations Management, 30*(1–2), 99–115.

Sezen, B. (2008). Relative effects of design, integration and information sharing on supply chain performance. *Supply Chain Management: An International Journal, 13*(3), 233–240.

Shah, S. A. A., Jajja, M. S. S., Chatha, K. A., & Farooq, S. (2020). Servitization and supply chain integration: An empirical analysis. *International Journal of Production Economics, 229*, 107765.

Shashi, Cerchione, R., Singh, R., Centobelli, P., & Shabani, A. (2018). Food cold chain management: From a structured literature review to a conceptual framework and research agenda. *International Journal of Logistics Management, 29*(3), 792–821.

Shee, H., Miah, S. J., Fairfield, L., & Pujawan, N. (2018). The impact of cloud-enabled process integration on supply chain performance and firm sustainability: The moderating role of top management. *Supply Chain Management: An International Journal, 23*(6), 500–517.

Shou, Y., Hu, W., Kang, M., Li, Y., & Park, Y. (2018). Risk management and firm performance: The moderating role of supplier integration. *Industrial Management & Data Systems, 118*(7), 1327–1344.

Shou, Y., Li, L., Kang, M., & Park, Y. (2020). Enhancing quality management through intra- and inter-plant integration in manufacturing networks. *Total Quality Management & Business Excellence, 31*(6), 623–635.

Shou, Y., Li, Y., Park, Y., & Kang, M. (2018). Supply chain integration and operational performance: The contingency effects of production systems. *Journal of Purchasing and Supply Management, 24*(4), 352–360.

Shou, Y., Li, Y., Park, Y. W., & Kang, M. (2017). The impact of product complexity and variety on supply chain integration. *International Journal of Physical Distribution & Logistics Management, 47*(4), 297–317.

Shou, Y., Shao, J., & Wang, W. (2021). How does reverse factoring affect operating performance? An event study of Chinese manufacturing firms. *International Journal of Operations & Production Management, 41*(4), 289–312.

Silvestro, R., & Lustrato, P. (2014). Integrating financial and physical supply chains: The role of banks in enabling supply chain integration. *International Journal of Operations & Production Management, 34*(3), 298–324.

Sinha, K. K., & Van de Ven, A. H. (2005). Designing work within and between organizations. *Organization Science, 16*(4), 389–408.

Srinivasan, R., & Swink, M. (2015). Leveraging supply chain integration through planning comprehensiveness: An organizational information processing theory perspective. *Decision Sciences, 46*(5), 823–861.

Stolze, H. J., Mollenkopf, D. A., Thornton, L., Brusco, M. J., & Flint, D. J. (2018). Supply chain and marketing integration: Tension in frontline social networks. *Journal of Supply Chain Management, 54*(3), 3–21.

Swink, M., Narasimhan, R., & Wang, C. (2007). Managing beyond the factory walls: Effects of four types of strategic integration on manufacturing plant performance. *Journal of Operations Management, 25*(1), 148–164.

Tarifa-Fernandez, J., de-Burgos-Jimenez, J., & Cespedes-Lorente, J. (2019). Absorptive capacity as a confounder of the process of supply chain integration. *Business Process Management Journal, 25*(7), 1587–1611.

Terjesen, S., Patel, P. C., & Sanders, N. R. (2012). Managing differentiation-integration duality in supply chain integration. *Decision Sciences, 43*(2), 303–339.

Tranfield, D., Denyer, D., & Smart, P. (2003). Towards a methodology for developing evidence-informed management knowledge by means of systematic review. *British Journal of Management, 14*(3), 207–222.

Turkulainen, V., Kauppi, K., & Nermes, E. (2017). Institutional explanations missing link in operations management? Insights on supplier integration. *International Journal of Operations & Production Management, 37*(8), 1117–1140.

Tushman, M. L., & Nadler, D. A. (1978). Information processing as an integrating concept in organizational design. *Academy of Management Review, 3*(3), 613–624.

van der Vaart, T., & van Donk, D. P. (2008). A critical review of survey-based research in supply chain integration. *International Journal of Production Economics, 111*(1), 42–55.

Vanpoucke, E., Vereecke, A., & Muylle, S. (2017). Leveraging the impact of supply chain integration through information technology. *International Journal of Operations & Production Management, 37*(4), 510–530.

Vickery, S. K., Jayaram, J., Droge, C., & Calantone, R. (2003). The effects of an integrative supply chain strategy on customer service and financial performance: An analysis of direct versus indirect relationships. *Journal of Operations Management, 21*(5), 523–539.

Vickery, S. K., Koufteros, X., & Droge, C. (2013). Does product platform strategy mediate the effects of supply chain integration on performance? A dynamic capabilities perspective. *IEEE Transactions on Engineering Management, 60*(4), 750–762.

References

Wei, S., Ke, W., Liu, H., & Wei, K. K. (2020). Supply chain information integration and firm performance: Are explorative and exploitative IT capabilities complementary or substitutive? *Decision Sciences, 51*(3), 464–499.

Wiengarten, F., Li, H. S., Singh, P. J., & Fynes, B. (2019). Re-evaluating supply chain integration and firm performance: Linking operations strategy to supply chain strategy. *Supply Chain Management: An International Journal, 24*(4), 540–559.

Wiengarten, F., Pagell, M., Ahmed, M. U., & Gimenez, C. (2014). Do a country's logistical capabilities moderate the external integration performance relationship? *Journal of Operations Management, 32*(1–2), 51–63.

Wong, C. W. Y., Wong, C. Y., & Boon-itt, S. (2013). The combined effects of internal and external supply chain integration on product innovation. *International Journal of Production Economics, 146*(2), 566–574.

Wong, C. Y., Boon-itt, S., & Wong, C. W. Y. (2011). The contingency effects of environmental uncertainty on the relationship between supply chain integration and operational performance. *Journal of Operations Management, 29*(6), 604–615.

Wong, C. Y., Wong, C. W. Y., & Boon-itt, S. (2020). Effects of green supply chain integration and green innovation on environmental and cost performance. *International Journal of Production Research, 58*(15), 4589–4609.

Wong, W. P., Sinnandavar, C. M., & Soh, K. L. (2021). The relationship between supply environment, supply chain integration and operational performance: The role of business process in curbing opportunistic behaviour. *International Journal of Production Economics, 232*, 107966.

Yang, Y., Jia, F., & Xu, Z. (2019). Towards an integrated conceptual model of supply chain learning: An extended resource-based view. *Supply Chain Management: An International Journal, 24*(2), 189–214.

Yu, W., Jacobs, M. A., Salisbury, W. D., & Enns, H. (2013). The effects of supply chain integration on customer satisfaction and financial performance: An organizational learning perspective. *International Journal of Production Economics, 146*(1), 346–358.

Zhang, M., Lettice, F., Chan, H. K., & Hieu Thanh, N. (2018). Supplier integration and firm performance: The moderating effects of internal integration and trust. *Production Planning & Control, 29*(10), 802–813.

Zhao, X., Huo, B., Flynn, B. B., & Yeung, J. H. Y. (2008). The impact of power and relationship commitment on the integration between manufacturers and customers in a supply chain. *Journal of Operations Management, 26*(3), 368–388.

Zhao, X., Huo, B., Selen, W., & Yeung, J. H. Y. (2011). The impact of internal integration and relationship commitment on external integration. *Journal of Operations Management, 29*(1–2), 17–32.

Zhu, S. N., Song, J. H., Hazen, B. T., Lee, K., & Cegielski, C. (2018). How supply chain analytics enables operational supply chain transparency: An organizational information processing theory perspective. *International Journal of Physical Distribution & Logistics Management, 48*(1), 47–68.

Chapter 3
Product Complexity, Variety and Supply Chain Integration

Abstract The need for aligning products and processes has been emphasized in the operations strategy literature for a long time. Drawing on the product–process fit perspective, this study aims to examine the relationships between product-level characteristics (i.e., product complexity and product variety) and the three dimensions of supply chain integration (SCI) (i.e., internal, supplier and customer integration). Structural equation modeling is employed to test the proposed hypotheses using survey data. The results show that for high product complexity, firms tend to implement internal and supplier integration, while product complexity does not directly impact customer integration. Product variety is confirmed as being positively related to all three dimensions of SCI. This study contributes to the SCI literature by providing empirical evidence on the relationships between product complexity, variety, and SCI.

Keywords Supply chain integration · Product complexity · Product variety · Internal integration · Supplier integration · Customer integration

3.1 Introduction

Van Donk and van der Vaart (2004) point out that few studies attempt to explore "what it is exactly that explains the differences in integrative practices in the supply chain" (p. 108). Chen et al. (2009) also call for "research on integration drivers or antecedents" (p. 75). In response to these calls, this study seeks to explore the key antecedents that drive the implementation of three SCI dimensions (i.e., internal, supplier and customer integration) at the product level.

Increasing product volume to meet market demand used to be the most sought-after goal of manufacturers. However, in today's business environment, manufacturers are facing the trends of market globalization and technological advancement. Firms are encouraged to provide a growing mix of products tailored to customers' differential

---

This chapter is a revised version of the following journal paper:

Shou, Y., Li, Y., Park, Y. W., & Kang, M. (2017). The impact of product complexity and variety on supply chain integration. *International Journal of Physical Distribution & Logistics Management, 47*(4), 297–317.

© The Author(s), under exclusive license to Springer Nature Singapore Pte Ltd. 2022
Y. Shou et al., *Supply Chain Integration for Sustainable Advantages*,
https://doi.org/10.1007/978-981-16-9332-8_3

needs (Chhetri et al., 2021), implying a need for a wider product variety. However, the difficulties of manufacturing these products tend to increase due to the large number of product components and the extensive interactions among these components, which leads to a high degree of product complexity (Novak & Eppinger, 2001). Product complexity and variety in manufacturing have created numerous challenges for supply chain management, which aims to improve flexibility and responsiveness to customer demands in a timely and cost-efficient manner (Bode & Wagner, 2015; Inman & Blumenfeld, 2014; Jüttner & Maklan, 2011; Ramdas, 2003). This study, therefore, focuses on the impact of these two aspects of products, i.e., product complexity and product variety.

In the manufacturing industry, many real-life examples illustrate how companies design and manage supply chains according to their product complexity and product diversity. For example, with respect to product complexity, Boeing puts great effort into supplier integration. The development of large passenger aircraft is a complex project that costs many resources. Boeing has more than 5,400 suppliers in different countries and regions worldwide, with approximately 500,000 supporting services (Boeing Company, 2021). To ensure the stability and safety of the supply chain, Boeing has established strict and comprehensive standards and procedures for supplier selection and development. Suppliers will be trained and reformed by the Boeing Company, following specific standards and systems. Moreover, to strengthen the real-time management and control of the global supply chain and realize on-time production, the Boeing Company uses advanced information systems to maintain close contact with suppliers worldwide. Regarding product variety, Procter & Gamble (P&G), as one of the largest consumer goods companies in the world, has a wide range of products (Procter & Gamble, 2021). P&G's market achievement is inseparable from its continuous exploration and innovation in the supply chain, especially in customer integration. The essence of the famous "P&G–Wal-Mart model" is the collaborative management of both parties' supply chains. These two companies were connected through electronic data interchange (EDI). In this way, P&G can know the inventory of different P&G products in Wal-Mart logistics centers in real time, as well as terminal data such as sales and prices, and carry out inventory/restocking management, production and R&D plans promptly to prevent inventory backlogs or shortages. Meanwhile, Wal-Mart uses EDI to obtain information from P&G to manage product shelves and subsequent purchases timely.

The above two examples show how companies with different product types integrate their supply chains. To proceed further, this study discusses the relationship between product complexity and variety and the internal, supplier and customer aspects of integration from the product–process fit perspective, complemented with governance theory and knowledge-based view (KBV). According to the theories related to product–process fit, it is argued that the processes are typically designed or organized to match the product design in manufacturing firms to achieve the best performance (Hayes & Wheelwright, 1984; Schmenner & Swink, 1998). Drawing upon transaction cost theory (Williamson, 1981, 1985), several studies have discussed the governance view of SCI (Das et al., 2006; Ellram & Cooper, 2014). As one type of hybrid governance structure, SCI can aid in reducing the "costs of running the

3.1 Introduction   33

system" (Das et al., 2006) such as negotiation, coordination and monitoring costs, and can thus achieve the same advantages as vertical integration (Shou et al., 2018) in safeguarding specific assets, processing complex information, and adapting to uncertainty through familiarity and trust among supply chain partners. Especially in the case of products with a high degree of complexity and variety, transaction and coordination costs across the whole supply chain increase with supply chain risk, which may demand integration among internal functional departments and external suppliers and customers. Additionally, product complexity and variety call for the integration of diversified knowledge. Prior work using KBV has noted that an integrative structure can facilitate the transfer and creation of knowledge within and across organizations (Kogut & Zander, 1992). The use of governance theory complemented with KBV provides the necessary theoretical underpinnings to fully understand how product complexity and variety influence firms' decisions on SCI.

Overall, we aim to investigate the impact of product complexity and variety on the three dimensions of SCI, which can provide empirical knowledge on why firms implement internal, supplier and customer integration based on product characteristics. This study contributes to the existing literature on product–process fit, especially the product design–supply chain interface. Furthermore, the results provide insights for managers to strengthen internal, supplier, and customer integration for highly complex and/or diversified products.

The rest of this chapter is organized as follows. In Sect. 3.2, we review the research on product–process fit and then develop the hypotheses that product complexity and variety will boost the implementation of all three dimensions of SCI. Section 3.3 introduces the data set and constructs the measurement of variables in the conceptual model. In Sect. 3.4, we apply structural equation modeling (SEM) to test the proposed hypotheses and reports the results. In Sect. 3.5, we discuss the major findings and implications of this study. Finally, Sect. 3.6 draws conclusions.

3.2 Literature Review and Hypotheses Development

3.2.1 Product–Process Fit

Many different conceptualizations of fit appear in the organization research and operations strategy literature (e.g., Chadwick et al., 2015; Drazin & Van de Ven, 1985; Mintzberg, 1979). For configurational theories, Mintzberg (1979) proposed that to survive or be maximally effective, organizations must design configurations that are internally consistent and fit multiple contextual dimensions, such as the organization's culture, environment, technology, size, age, or tasks. Resource orchestration theory holds that for a firm, the fit and combination of resources, capabilities, and management practices could determine firm performance (Chadwick et al., 2015). Drazin and Van de Ven (1985) summarized many of these conceptualizations into the interaction, selection, and systems approaches to fit. The interaction approach to fit characterizes many traditional theories, especially "contingency" theories, which define fit as the interaction of two variables (Venkatraman, 1989). The selection approach to fit is adopted by theorists who develop organizational taxonomies (Doty et al., 1993). The systems approach defines fit in terms of consistency across multiple dimensions of organizational design and the context. Among these three approaches, the product–process fit is consistent with the interaction approach.

Discussion of the fit between product and process can be found primarily in the operations strategy literature. The product–process matrix proposed by Hayes and Wheelwright (1979) has wide acceptance among these studies. According to the understanding of the follow-up research, its central proposition is that product and process should be matched for manufacturing plants to achieve superior operational performance (Hayes & Wheelwright, 1984; Schmenner & Swink, 1998). Essentially, the product–process matrix defines the product variety and volume that is most suitable for the four typical process types (i.e., job shop, batch shop, assembly line, and continuous flow line). Production with low volume and low standardization should be organized as a jumbled flow (job shop), moderate-variety and low-volume production as a disconnected line flow (batch shop), low-variety and high-volume production as a connected line flow (assembly line), and commodity production as a continuous flow line (Hayes & Wheelwright, 1984). Empirical evidence has shown that many manufacturers indeed adopted the proposition of this product–process matrix (Das & Narasimhan, 2001). As a result, many follow-up studies have adopted this matrix.

However, based on the studies that tested the product–process matrix, Helkiö and Tenhiälä (2013) have found significant deviations from the proposition of the original product–process matrix. In the aspect of the product, the dimensions (i.e., variety and volume) of the matrix are too narrow. For example, Kotha and Orne (1989) have suggested that the complexity of products is also a critical influencing factor for process design. In the aspect of the process, characteristics of the four typical processes are changing in modern industrial reality. For example, in the product–process matrix, job shops represent flexible processes, and assembly lines and continuous flow lines represent inflexible processes (Lummus et al., 2006). However, flexible manufacturing systems (FMSs) have enabled wide-variety production in assembly lines and flow lines (Helkiö & Tenhiälä, 2013).

Nowadays, the changing production context has extended the principles of product–process fit for production, service, and combined product–service offerings (Buzacott, 2000; Hullova et al., 2016). For example, in today's highly competitive global business landscape, customers demand personalized products and responsive distribution systems; hence Kumar et al. (2020) developed a distributed manufacturing strategy based on the product–process matrix. Moreover, advanced manufacturing and information technologies have also affected product–process fit. Digital transformation will impact the adaptive capacity of the manufacturing firm to adjust its business model accordingly and reflect customer demand patterns (Weller et al., 2015). For example, Eyers et al. (2021) conducted a case study to investigate the impact of three-dimensional (3D) printing on the original product–process matrix,

3.2 Literature Review and Hypotheses Development 35

and found that 3D printing can help overcome some traditional constraints. In this present study, we attempt to extend the product–process fit concept and investigate the impact of product complexity and variety at the supply chain level.

3.2.2 Product Complexity and SCI

Within the area of operations management, the concept of product complexity has been associated with "the number of parts or components needed to build the product" (Inman & Blumenfeld, 2014, p. 1957). Lucchetta et al. (2005) and Kaufmann and Carter (2006) also define complexity from a technical perspective as the difficulty of generating or manufacturing parts or components. Bode and Wagner (2015) have summarized these different definitions in terms of two aspects: structural complexity (number and variety of elements) and operational complexity (interactions between elements). This study adopts this definition of Bode and Wagner (2015), since structural and operational dimensions present a more comprehensive view of product complexity. Inman and Blumenfeld (2014) consider product complexity as one of the critical risk factors which further influence supply chain strategy. Similarly, this study assumes that manufacturers of complex products need to strengthen SCI implementation to govern risk factors incurred by product complexity.

Internal integration is encouraged in manufacturing firms with high levels of product complexity (Kotha & Orne, 1989). One reason for this is that complex products with multiple components are strongly associated with difficulties in design and production (Chhetri et al., 2021), thereby increasing transaction and coordination costs between functional departments. In this case, the purchasing, manufacturing, and marketing departments within the firm must work closely together to support concurrent engineering and design for manufacturing (Swink & Nair, 2007). Moreover, in environments with high product complexity, manufacturers are required to deal with component inventory and capacity problems (Eckstein et al., 2015), which results in frequent coordination and collaboration between manufacturing and purchasing departments.

While most researchers have acknowledged the significant effects of product complexity on the internal manufacturing strategy of the firm, few discuss its specific impact on external supply chain strategy. A higher level of product complexity is related to higher supply chain risks and disruptions (Bode & Wagner, 2015; Inman & Blumenfeld, 2014), which increase difficulties in coordinating supply and demand. A complex product may contain components or parts that each have different technical specifications and lead times (Lu et al., 2003). The more complex the final product is, the more difficult it is to specify all specifications and production schedules (Kaufmann & Carter, 2006). If the production and delivery of a particular component of a complex product experience difficulties or delays, it is likely to increase costs substantially. In general, a dual or even multiple sourcing strategies or a buffering policy are recommended for dealing with these disruption risks (Sawik, 2013). However, for complex products, these approaches will also increase the complexity of the supply chain network as a whole, since they require high levels of coordination with greater numbers of suppliers, and involve more customer approval processes for using the sourced components in production. We, therefore, infer that for high levels of product complexity, transaction and coordination costs in the exchanges between manufacturers and their external supply chain partners increase, which calls for close and collaborative relationship.

Specifically, the integration that includes information sharing, adequate coordination and collaboration with key external partners is regarded as effective in preventing and eliminating the uncertainty and supply chain risks arising from product complexity (Ambulkar et al., 2015). Customer integration enables manufacturers to attain accurate information on demand in order to specify the quality and quantity requirements of products in detail (Flynn et al., 2010). Manufacturers also transmit this information to suppliers, which improves the suppliers' understanding and anticipation of the manufacturer's needs (Flynn et al., 2010). The sharing of demand information is also beneficial for suppliers in arranging production and inventory of the parts or components. Therefore, integration with customers and suppliers increases the transparency of the complex information about the product and the production process in the supply chain. Moreover, collaborative approaches such as the sharing of benefits and risks increase the willingness of supply chain partners to exchange critical information and knowledge, thus guaranteeing the delivery of components. We conclude that external integration with suppliers and customers through the breaking-down of organizational walls is necessary to cope with the negative effects of product complexity.

Furthermore, KBV supports the governance view in the selection of an integrative structure for firms manufacturing complex products. Kaufmann and Carter (2006) point out that product complexity requires close cooperation between manufacturers and their external partners to achieve the benefits of joint actions, especially in the early stages of product development. KBV suggests that the frequencies and values of knowledge transfer and creation are both much higher in the context of solving a complex problem (Nickerson & Zenger, 2004). Despite the difficulty of knowledge transfer across organizational boundaries (Kogut & Zander, 1992), manufacturers can build a set of organizational structures, rules, principles, routines, channels, and procedures through the unified control of supply chain processes and activities to promote the transfer, convergence, and creation of knowledge from multiple organizations. When a manufacturing firm achieves this level of integration with its external partners, the supply chain acts as a social community "in which individual and social expertise is transformed into economically useful products and services by the application of a set of higher-order organizing principles" (Kogut & Zander, 1992, p. 384).

In short, from the perspective of product–process fit, as a critical product-level characteristic, product complexity may demand the support of internal, supplier, and customer integration to achieve superior firm performance. Therefore, we propose that:

3.2 Literature Review and Hypotheses Development

H1a The higher the product complexity, the more likely that manufacturing firms promote internal integration.

H1b The higher the product complexity, the more likely that manufacturing firms promote supplier integration.

H1c The higher the product complexity, the more likely that manufacturing firms promote customer integration.

3.2.3 Product Variety and SCI

Product variety is defined by Fisher et al. (1999) as "the breadth of products that a firm offers at a given time" (p. 197). Randall and Ulrich (2001) refer to this as "the number of different versions of a product offered by a firm at a single point in time" (p. 1588). The definition of Fisher et al. (1999) is applied here, since this study focuses on the variety of product portfolios in the manufacturing firm. Several scholars have associated product variety with product innovation or new product development (NPD) activities. Indeed, "increasing product variety implies that new products are introduced" (Wan et al., 2012, p. 318). In this study, we posit that SCI is encouraged by a manufacturer with a variety of product portfolios.

In the product–process matrix of Hayes and Wheelwright (1979), product variety is a primary product characteristic in addition to product volume, which influences the degree of flexibility of process structure. It can be inferred that a high level of product variety gives rise to a widespread of product and production information in the supply chain. Excessive product information may result in selection confusion (variety fatigue) for customers and lead to forecasting difficulty for manufacturers (Um et al., 2017; Wan et al., 2012). Meanwhile, variations in product configuration present great difficulties for the manufacturer in terms of coordinating suppliers. In this case, the manufacturer not only needs to process large amounts of product design information to cater to customers' changing needs, but also needs to manage the production information for various components and modules in order to coordinate its upstream supply, which greatly increases transaction and coordination costs. We argue that there is a greater need for SCI to implement a strategy of product variety effectively.

Specifically, internal integration facilitates the transfer and recombination of ideas, knowledge, and information dispersed across functional departments, and this is beneficial for the firm in building product portfolios that are robust against environmental changes (Um et al., 2017). The firm can also improve its information-processing capability through internal integration (Wong et al., 2011). Integration with customers can enrich manufacturers' knowledge of product demands, requirements, and market trends (Flynn et al., 2010), which helps manufacturers grasp opportunities and develop competitive strengths. Additionally, the difficulties in scheduling production resulting from product variety also call for information sharing and coordinated actions with suppliers (Randall & Ulrich, 2001; Wan et al., 2012). Extensive supplier integration is encouraged to ensure that suppliers deliver components or modules in a timely and accurate manner. Moreover, since suppliers often control vitally important new technology or knowledge about components or materials for certain types of products (Arya et al., 2008), integration offers a more efficient way to access this knowledge than traditional market relationships.

KBV can also be used to explain why a firm implements an integrative structure for high levels of product variety. In order to continuously introduce new products, a product diversification strategy requires the integration of both internal and external complementary knowledge (and especially implicit knowledge) across the different value chain activities and organizations (Al Zu'Bi & Tsinopoulos, 2012) to improve the firm's innovation capability. KBV suggests that the efficiency of transferring complementary knowledge depends on the level of authority in directing others' actions (Conner & Prahalad, 1996) and the formation of a shared language and identity (Kogut & Zander, 1996), which motivates firms to engage in collaborative arrangements within and across organizations (Grant, 1996; Lavie et al., 2010). As Schoenherr and Swink (2012) note, "to develop such organizational skills (to acquire and exploit unique knowledge), a firm typically must work on creating effective communication protocols, shared understandings and languages, and shared collaborative values with supply chain partners" (p. 101).

We therefore propose the second hypothesis as follows:

H2a The higher the level of product variety, the more likely that manufacturing firms promote internal integration.

H2b The higher the level of product variety, the more likely that manufacturing firms promote supplier integration.

H2c The higher the level of product variety, the more likely that manufacturing firms promote customer integration.

3.3 Method

3.3.1 Data

To test the theoretical hypotheses proposed above, data collected from the sixth round International Manufacturing Strategy Survey (IMSS) were used. More details about the IMSS dataset are available in Chap. 1 of this book (Shou et al., 2022). The final sample used in this study has 843 valid responses after dropping responses with missing data.

3.3.2 Measures

All measures for product complexity, product variety, and SCI align with the existing literature. Detailed survey questions are available in the Appendix of this book (Shou et al., 2022).

There are two independent variables, i.e., product complexity and product variety. The three items of integrated design, complexity of bill of material and number of operational steps required to build the plant's products are used to measure product complexity (Inman & Blumenfeld, 2014; Lucchetta et al., 2005). Product variety is defined as the range of products offered by the plants (Fisher et al., 1999). In general, firms diversify product versions by frequently introducing innovative products (Miller & Roth, 1994; Wan et al., 2012). Therefore, respondents were required to assess the attributes of variety with regard to a wider product range (Fisher et al., 1999), offering new products more frequently (Wan et al., 2012) and offering more innovative products (Frohlich & Dixon, 2001; Miller & Roth, 1994).

This study characterizes external integration in terms of collaborative approaches, information sharing, joint decision-making on contingencies and system coupling (Ellinger et al., 2000; Flynn et al., 2010; Narasimhan & Kim, 2002; Spekman, 1988). Customer integration and supplier integration were each measured using four items ranging from one (none) to five (high) indicating the current level of implementation. As for internal integration, information sharing and joint decision making are recognized by Ellinger et al. (2000), Narasimhan and Kim (2002), and Huo et al. (2015) and as being important for internal integration in order to coordinate production and inventory management and ensure customer service. In this study, internal integration was therefore measured in terms of joint decision making and information sharing between purchasing and sales departments.

In addition to the product-level characteristics which may influence a firm's decision on SCI, we control for firm size in terms of the natural logarithm of the total number of employees. Firm size is often used as a control variable for two reasons, one of which is that larger firms tend to have more resources and capabilities for carrying out activities (Kimberly, 1976); the other is that larger firms can "take advantages of economies of scale in their business activities" (Kim & Lee, 2010, p. 964).

3.3.3 Reliability and Validity

To evaluate the focal constructs, exploratory factor analysis (EFA) was first carried out. A principal component analysis with Varimax rotation was conducted to test the unidimensionality of each construct and Cronbach's alpha was calculated to assess the internal consistency of the construct (Cronbach, 1951). The EFA results revealed that all items had strong loadings on the construct that they were intended to measure and lower loadings on other constructs, thus confirming construct unidimensionality.

As recommended by Nunnally et al. (1978) and Fornell and Larcker (1981), Cronbach's alpha for each construct was greater than 0.60. The reliability of the five constructs is therefore ensured.

Confirmatory factor analysis (CFA) with the maximum likelihood method was used to examine convergent validity and discriminant validity. The ratio of χ^2 to the degree of freedom (χ^2/df) was 3.873. Since "a large sample size may cause the rejection of almost any model, even for models that explain most of the variance in the data" (O'Leary-Kelly & Vokurka, 1998, p. 403), other fit indices should therefore be examined for thoroughness of discussion. According to Hu and Bentler (1999), the fit indices of RMSEA = 0.058, SRMR = 0.046, CFI = 0.953, GFI = 0.940, IFI = 0.953 and TLI = 0.942 indicate a reasonably high level of fit for the model. As summarized in Table 3.1, most of the factor loadings in the CFA model, except for PC1 (0.459), are greater than 0.50, with the smallest t-value being 11.941. Nonetheless, this item is retained since PC1 "integrated product design" is used to describe the operational dimension (i.e., interactions between the elements) of product complexity (Inman & Blumenfeld, 2014; Lucchetta et al., 2005). In addition, the EFA results show that PC1 has a factor loading of greater than 0.50 (0.683), and is significantly loaded on product complexity. Product complexity has a relatively low average variance extracted (AVE = 0.45). Values for AVE for the other four constructs are all higher than 0.50, and the composite reliability (CR) for all five constructs are greater than

Table 3.1 Factor analysis results

Construct	Item	Factor loading	S.E	t-value	Reliability and validity
Product complexity	PC1	0.459	0.044	11.941	Cronbach's alpha = 0.677; CR = 0.702; AVE = 0.454
	PC2	0.663	0.046	16.611	
	PC3	0.844	0.044	19.652	
Product variety	PV1	0.595	0.034	17.088	Cronbach's alpha = 0.757; CR = 0.770; AVE = 0.533
	PV2	0.868	0.038	24.968	
	PV3	0.701	0.036	20.166	
Internal integration	II1	0.753	0.030	21.884	Cronbach's alpha = 0.887; CR = 0.887; AVE = 0.664
	II2	0.784	0.031	23.609	
	II3	0.855	0.028	30.736	
	II4	0.862	0.030	31.492	
Supplier integration	SI1	0.768	0.030	24.996	Cronbach's alpha = 0.842; CR = 0.850; AVE = 0.587
	SI2	0.822	0.030	27.752	
	SI3	0.789	0.032	26.171	
	SI4	0.677	0.037	21.143	
Customer integration	CI1	0.853	0.032	29.818	Cronbach's alpha = 0.881; CR = 0.884; AVE = 0.657
	CI2	0.854	0.033	29.864	
	CI3	0.753	0.034	24.792	
	CI4	0.776	0.037	25.817	

3.3 Method

Table 3.2 Correlations of the constructs

	(1)	(2)	(3)	(4)	(5)
(1) Product complexity	0.674				
(2) Product variety	0.199[***]	0.730			
(3) Internal integration	0.228[***]	0.233[***]	0.815		
(4) Supplier integration	0.284[***]	0.345[***]	0.641[***]	0.766	
(5) Customer integration	0.187[***]	0.340[***]	0.531[***]	0.746[***]	0.810

Note Value on the diagonal is the square-root of AVE. [***]$p < 0.001$

0.70. This demonstrates the reliability of the measurement scales (Fornell & Larcker, 1981; O'Leary-Kelly & Vokurka, 1998). From these results, convergent validity is deemed to be ensured.

Referring to Fornell and Larcker (1981), the measures have good discriminant validity when the square root of the AVE for each construct is greater than its correlation with other constructs. The results presented in Table 3.2 verify a satisfactory level of discriminant validity. We also test the discriminant validity by comparing the unconstrained model (with the two constructs allowed to vary freely) with the constrained model (with the correlations between two constructs constrained to 1) (Bagozzi et al., 1991; O'Leary-Kelly & Vokurka, 1998). Significant differences in χ^2 provide further evidence for discriminant validity.

3.4 Results

As a widely-used technique for simultaneous testing of the complex, multi-stage relationships between variables (Lomax & Schumacker, 2012), structural equation modeling (SEM) was used with the maximum likelihood estimation method to test the proposed hypotheses. The goodness of fit indices for our model are $\chi^2/df =$ 3.667, RMSEA $= 0.056$, 90% confidence interval for RMSEA $= (0.051; 0.062)$, GFI $= 0.940$, NFI $= 0.935$, NNFI (TLI) $= 0.940$, IFI $= 0.952$, CFI $= 0.952$, RFI $=$ 0.920, standardized RMR $= 0.045$, AGFI $= 0.917$, PGFI $= 0.683$. These indices are all better than the recommended thresholds (Hu & Bentler, 1999), which indicates that the overall fits of the model are good. Figure 3.1 and Table 3.3 show the fit indices and the results of hypotheses testing.

H1 examines the relationship between product complexity and (a) internal integration, (b) supplier integration and (c) customer integration. H1a is supported with a path coefficient of 0.190 ($t = 4.468$), which is statistically significant at the level of 0.001. The path coefficient for H1b supplier integration is 0.104 ($t = 2.816$), which is statistically significant at the level of 0.01. These results confirm that product complexity has significant, positive, and direct impacts on internal integration and supplier integration. However, the path coefficient for H1c (customer integration) is

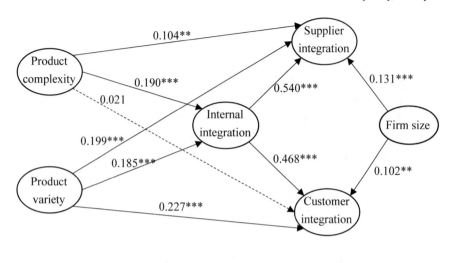

^{2}/df =3.667, RMSEA = 0.056, SRMR = 0.045, CFI = 0.952, NNFI (TLI) = 0.940
***p<0.001; **p<0.01

———————▶ significant path;   - - - - - - -▶ non-significant path

Fig. 3.1 Estimated structural equation model

Table 3.3 SEM path analysis

	Structural paths	Standardized estimates	S.E	p-value	t-value
H1a	Product complexity → internal integration	0.190***	0.032	0.000	4.468
H1b	Product complexity → supplier integration	0.104**	0.035	0.005	2.816
H1c	Product complexity → customer integration	0.021	0.037	0.574	0.564
H2a	Product variety → internal integration	0.185***	0.037	0.000	4.467
H2b	Product variety → supplier integration	0.199***	0.041	0.000	5.502
H2c	Product variety → customer integration	0.227***	0.044	0.000	6.011

***p < 0.001; **p < 0.01

not statistically significant, with a value of 0.021 ($t = 0.564$). H1c therefore cannot be confirmed.

H2 predicts that product variety is positively associated with internal integration, supplier integration and customer integration. The standardized path coefficients are 0.185 ($t = 4.467$), 0.199 ($t = 5.502$) and 0.227 ($t = 6.011$), respectively. The coefficients for H2 are all statistically significant at the level of 0.001, which indicates

3.4 Results

that product variety has significant, positive, and direct effects on internal integration, supplier integration and customer integration. Thus, H2 is supported.

Furthermore, we calculate the indirect effects of product complexity and variety on SCI to determine whether internal integration carries the effects of product complexity and variety to external integration. By multiplying the path coefficients from product complexity/variety to internal integration and from internal integration to customer/supplier integration, the indirect effect of product complexity on customer integration is shown to be $0.190 \times 0.468 = 0.089$, that of product complexity on supplier integration is $0.190 \times 0.540 = 0.103$, that of product variety on customer integration is $0.185 \times 0.468 = 0.087$, and that of product variety on supplier integration is $0.185 \times 0.540 = 0.100$. To test the significance of these indirect effects, Sobel's Z-test was conducted, and the resultant Z-values for the above indirect effects are 4.219 ($p = 0.000$), 4.291 ($p = 0.000$), 4.117 ($p = 0.000$), and 4.184 ($p = 0.000$), respectively. These results indicate that internal integration is a mediating factor between product complexity/variety and supplier/customer integration.

3.5 Discussion

3.5.1 Findings

This study evidences that product complexity has direct and positive effects on internal integration and supplier integration whereas there is no significant relationship between product complexity and customer integration. Higher levels of product complexity often accompany higher supply chain risks, and problems that happen in the production or delivery of a single component may result in a massive waste of resources. Information sharing and cooperation with supply chain partners can effectively reduce the waste of resources caused by product complexity. Our result indicates that product complexity tends to push manufacturers to integrate with suppliers rather than consumers. This result is consistent with the example of the Boeing Company we mentioned earlier. Although Boeing's supply chain is very complicated, it is mainly at the supplier's side, and hence the company does not have to invest too many resources in customer integration.

Meanwhile, product variety has direct and positive effects on internal, supplier, and customer integration. This result shows that for all levels of product complexity or product diversity, supplier integration is essential for manufacturing firms. Companies that produce diversified products need to have accurate sales forecasts and real-time information, so they need more information exchange and cooperation with customers. These results illustrate that supply chain practices differ between organizations and each firm is recommended to implement key SCI dimensions tactically.

In addition, product complexity and variety indirectly influence supplier and customer integration through internal integration, which indirectly illustrates the priority of internal integration in all the three SCI dimensions. This result is consistent with the classic studies from the product–process fit perspective.

3.5.2 Theoretical Implications

According to the product–process fit literature, to achieve superior performance, the process should be organized or designed for product design in manufacturing firms, which can be further confirmed and extended in this study. Moreover, the original "process" in classic studies is defined as the layout of the production process. We attempt to extend the process idea to the supply chain. The results of this study illustrate that product–process fit needs to be realized not only within manufacturing firms, but also among the supply chain partners.

Moreover, the results help answer the question of why a manufacturer implements internal integration, supplier integration, and customer integration at the product level; these results are presented in response to the call made by Van Donk and van der Vaart (2004) and Chen et al. (2009) for further research on the antecedents of SCI. Combining the governance view and KBV, we conclude that product complexity and variety give rise to a need for SCI to mitigate transactional hazards and facilitate knowledge transfer.

3.6 Conclusions

In summary, this research focuses on the impact of product characteristics on the implementation of internal, supplier and customer integration. The empirical results show that product complexity and variety are two critical antecedents for the implementation of SCI. This research extends the literature on SCI and contributes to discussion of product design–supply chain interface. We also find that firms have a particular need to conduct SCI in the case of high product complexity and variety.

Although this study offers new insights into SCI literature and practice, there are some limitations to this study. First, cross-sectional data were used to test the proposed hypotheses. Since firms with high levels of SCI may have advantages in offering complex and diverse products, longitudinal studies will be helpful in revealing the evolutional patterns of product design and SCI. Second, further scrutiny of supply chain complexity is required. In this study, only complexity and variety at the product level were considered. It therefore would be valuable for future research to investigate the impact of other potential sources of supply chain complexity on SCI. Finally, this study focused solely on the antecedents of SCI at the product level. Since SCI plays an important role in handling product complexity and variety and in facilitating knowledge transfer and creation, a further examination of the ways in which SCI influences product innovation performance in the context of product complexity and variety would be valuable.

3.6 Conclusions

Acknowledgements This work was supported by the National Natural Science Foundation of China under Grant Number 71472166, the Fundamental Research Funds of Shandong University, and Shandong Provincial Social Science Foundation under Grant Number 16CGLJ13.

References

Al Zu'Bi, Z. M., & Tsinopoulos, C. (2012). Suppliers versus lead users: Examining their relative impact on product variety. *Journal of Product Innovation Management, 29*(4), 667–680.

Ambulkar, S., Blackhurst, J., & Grawe, S. (2015). Firm's resilience to supply chain disruptions: Scale development and empirical examination. *Journal of Operations Management, 33–34*, 111–122.

Arya, A., Mittendorf, B., & Sappington, D. E. (2008). Outsourcing, vertical integration, and price vs. quantity competition. *International Journal of Industrial Organization, 26*(1), 1–16.

Bagozzi, R. P., Yi, Y., & Phillips, L. W. (1991). Assessing construct validity in organizational research. *Administrative Science Quarterly, 36*(3), 421–458.

Bode, C., & Wagner, S. M. (2015). Structural drivers of upstream supply chain complexity and the frequency of supply chain disruptions. *Journal of Operations Management, 3*, 215–228.

Boeing Company. (2021). *Boeing around the globe.* https://www.boeing.com/global/. Accessed 14 Sept 2021.

Buzacott, J. A. (2000). Service system structure. *International Journal of Production Economics, 68*(1), 15–27.

Chadwick, C., Super, J. F., & Kwon, K. (2015). Resource orchestration in practice: CEO emphasis on SHRM, commitment-based HR systems, and firm performance. *Strategy Management Journal, 36*, 360–376.

Chen, H., Daugherty, P. J., & Roath, A. S. (2009). Defining and operationalizing supply chain process integration. *Journal of Business Logistics, 30*(1), 63–84.

Chhetri, P., Hashemi, A., Lau, K. H., & Lim, M. K. (2021). Aligning supply chain complexity with product demand and design characteristics. *International Journal of Logistics: Research and Applications*, 1–27.

Conner, K. R., & Prahalad, C. K. (1996). A resource-based theory of the firm: Knowledge versus opportunism. *Organization Science, 7*(5), 477–501.

Cronbach, L. J. (1951). Coefficient alpha and the internal structure of tests. *Psychometrika, 16*(3), 297–334.

Das, A., & Narasimhan, R. (2001). Process-technology fit and its implications for manufacturing performance. *Journal of Operations Management, 19*(5), 521–540.

Das, A., Narasimhan, R., & Talluri, S. (2006). Supplier integration—Finding an optimal configuration. *Journal of Operations Management, 24*(5), 563–582.

Doty, H. D., Glick, W. H., & Huber, G. P. (1993). Fit, equifinality, and organizational effectiveness: A test of two configurational theories. *Academy of Management Journal, 36*(6), 1196–1250.

Drazin, R., & Van de Ven, A. H. (1985). Alternate forms of fit in contingency theory. *Administrative Science Quarterly, 30*, 514–539.

Eckstein, D., Goellner, M., Blome, C., & Henke, M. (2015). The performance impact of supply chain agility and supply chain adaptability: The moderating effect of product complexity. *International Journal of Production Research, 53*(10), 3028–3046.

Ellinger, A. E., Daugherty, P. J., & Keller, S. (2000). The relationship between marketing/logistics interdepartmental integration and performance in US manufacturing firms: An empirical study. *Journal of Business Logistics, 21*(1), 1–22.

Ellram, L. M., & Cooper, M. C. (2014). Supply chain management: It's all about the journey, not the destination. *Journal of Supply Chain Management, 50*(1), 8–20.

Eyers, D. R., Potter, A. T., Gosling, J., & Naim, M. M. (2021). The impact of additive manufacturing on the product-process matrix. *Production Planning & Control, 1–17.* https://doi.org/10.1080/09537287.2021.1876940

Fisher, M., Ramdas, K., & Ulrich, K. (1999). Component sharing in the management of product variety: A study of automotive braking systems. *Management Science, 45*(3), 297–315.

Flynn, B. B., Huo, B., & Zhao, X. (2010). The impact of supply chain integration on performance: A contingency and configuration approach. *Journal of Operations Management, 28*(1), 58–71.

Fornell, C., & Larcker, D. F. (1981). Evaluating structural equation models with unobservable variables and measurement error. *Journal of Marketing Research, 18*(1), 39–50.

Frohlich, M. T., & Dixon, J. R. (2001). A taxonomy of manufacturing strategies revisited. *Journal of Operations Management, 19*(5), 541–558.

Grant, R. M. (1996). Toward a knowledge-based theory of the firm. *Strategic Management Journal, 17,* 109–122.

Hayes, R. H., & Wheelwright, S. C. (1979). Link manufacturing process and product life cycles. *Harvard Business Review, 57*(1), 133–140.

Hayes, R. H., & Wheelwright, S. C. (1984). *Restoring our competitive edge: Competing through manufacturing.* Wiley.

Helkiö, P., & Tenhiälä, A. (2013). A contingency theoretical perspective to the product-process matrix. *International Journal of Operations & Production Management, 33*(2), 216–244.

Hu, L. T., & Bentler, P. M. (1999). Cutoff criteria for fit indexes in covariance structure analysis: Conventional criteria versus new alternatives. *Structural Equation Modeling, 6*(1), 1–55.

Hullova, D., Trott, P., & Simms, C. D. (2016). Uncovering the reciprocal complementarity between product and process innovation. *Research Policy, 45*(5), 929–940.

Huo, B., Han, Z., Chen, H., & Zhao, X. (2015). The effect of high-involvement human resource management practices on supply chain integration. *International Journal of Physical Distribution & Logistics Management, 45*(8), 716–746.

Inman, R. R., & Blumenfeld, D. E. (2014). Product complexity and supply chain design. *International Journal of Production Research, 52*(7), 1956–1969.

Jüttner, U., & Maklan, S. (2011). Supply chain resilience in the global financial crisis: An empirical study. *Supply Chain Management: An International Journal, 16*(4).

Kaufmann, L., & Carter, C. R. (2006). International supply relationships and non-financial performance—A comparison of US and German practices. *Journal of Operations Management, 24*(5), 653–675.

Kim, D., & Lee, R. P. (2010). Systems collaboration and strategic collaboration: Their impacts on supply chain responsiveness and market performance. *Decision Sciences, 41*(4), 955–981.

Kimberly, J. R. (1976). Organizational size and the structuralist perspective: A review, critique, and proposal. *Administrative Science Quarterly, 21*(4), 571–597.

Kogut, B., & Zander, U. (1992). Knowledge of the firm, combinative capabilities, and the replication of technology. *Organization Science, 3*(3), 383–397.

Kogut, B., & Zander, U. (1996). What firms do? Coordination, identity, and learning. *Organization Science, 7*(5), 502–518.

Kotha, S., & Orne, D. (1989). Generic manufacturing strategies: A conceptual synthesis. *Strategic Management Journal, 10*(3), 211–231.

Kumar, M., Tsolakis, N., Agarwal, A., & Srai, J. S. (2020). Developing distributed manufacturing strategies from the perspective of a product-process matrix. *International Journal of Production Economics, 219,* 1–17.

Lavie, D., Stettner, U., & Tushman, M. L. (2010). Exploration and exploitation within and across organizations. *Academy of Management Annals, 4*(1).

Lomax, R. G., & Schumacker, R. E. (2012). *A beginner's guide to structural equation modeling.* Routledge.

Lu, Y., Song, J., & Yao, D. D. (2003). Order fill rate, leadtime variability, and advance demand information in an assemble-to-order system. *Operations Research, 51*(2), 292–308.

References

Lucchetta, G., Bariani, P. F., & Knight, W. A. (2005). Integrated design analysis for product simplification. *CIRP Annals-Manufacturing Technology, 54*(1), 147–150.

Lummus, R. R., Vokurka, R. J., & Duclos, L. K. (2006). The produce-process matrix revisited: Integrating supply chain trade-offs. *SAM Advanced Management Journal, 71*(2), 4–45.

Miller, J. G., & Roth, A. V. (1994). A taxonomy of manufacturing strategies. *Management Science, 40*(3), 285–304.

Mintzberg, H. T. (1979). *The structuring of organizations*. Prentice-Hall.

Narasimhan, R., & Kim, S. W. (2002). Effect of supply chain integration on the relationship between diversification and performance: Evidence from Japanese and Korean firms. *Journal of Operations Management, 20*(3), 303–323.

Nickerson, J. A., & Zenger, T. R. (2004). A knowledge-based theory of the firm—The problem-solving perspective. *Organization Science, 15*(6), 617–632.

Novak, S., & Eppinger, S. D. (2001). Sourcing by design: Product complexity and the supply chain. *Management Science, 24*(47), 189–204.

Nunnally, J. C., Bernstein, I. H., & Berge, J. M. T. (1978). *Psychometric theory*. McGraw-Hill.

O'Leary-Kelly, S. W., & Vokurka, R. J. (1998). The empirical assessment of construct validity. *Journal of Operations Management, 16*(4), 387–405.

Procter & Gamble. (2021). *P&G's supply chain*. https://www.pgcareers.com/supply-chain-masters. Accessed 14 Sept 2021.

Randall, T., & Ulrich, K. (2001). Product variety, supply chain structure, and firm performance: Analysis of the US bicycle industry. *Management Science, 47*(12), 1588–1604.

Ramdas, K. (2003). Managing product variety: An integrative review and research directions. *Production and Operations Management, 12*(1), 79–101.

Sawik, T. (2013). Selection of resilient supply portfolio under disruption risks. *Omega, 41*, 259–269.

Schmenner, R. W., & Swink, M. L. (1998). On theory in operations management. *Journal of Operations Management, 17*(1), 97–113.

Schoenherr, T., & Swink, M. (2012). Revisiting the arcs of integration: Cross-validations and extensions. *Journal of Operations Management, 30*(1), 99–115.

Shou, Y., Kang, M., & Park, Y. (2022). *Supply chain integration for sustainable advantages*. Springer.

Shou, Y., Li, Y., Park, Y., & Kang, M. (2018). Supply chain integration and operational performance: The contingency effects of production systems. *Journal of Purchasing and Supply Management, 24*(4), 352–360.

Spekman, R. E. (1988). Strategic supplier selection: Understanding long-term buyer relationships. *Business Horizons, 31*(4), 75–81.

Swink, M., & Nair, A. (2007). Capturing the competitive advantages of AMT: Design–manufacturing integration as a complementary asset. *Journal of Operations Management, 25*, 736–754.

Um, J., Lyons, A., Lam, H. K., Cheng, T. C. E., & Dominguez-Pery, C. (2017). Product variety management and supply chain performance: A capability perspective on their relationships and competitiveness implications. *International Journal of Production Economics, 187*, 15–26.

Van Donk, D. P., & van der Vaart, T. (2004). Business conditions, shared resources and integrative practices in the supply chain. *Journal of Purchasing and Supply Management, 10*(3), 107–116.

Venkatraman, N. (1989). The concept of fit in strategy research: Toward verbal and statistical correspondence. *Academy of Management Review, 14*(3), 423–444.

Wan, X., Evers, P. T., & Dresner, M. E. (2012). Too much of a good thing: The impact of product variety on operations and sales performance. *Journal of Operations Management, 30*(4), 316–324.

Weller, C., Kleer, R., & Piller, F. T. (2015). Economic implications of 3D printing: Market structure models in light of additive manufacturing revisited. *International Journal of Production Economics, 164*, 43–56.

Williamson, O. E. (1981). The economics of organization: The transaction cost approach. *American Journal of Sociology, 87*(3), 548–577.

Williamson, O. E. (1985). *The economic institutions of capitalism*. Free Press.

Wong, C. Y., Boon-Itt, S., & Wong, C. W. (2011). The contingency effects of environmental uncertainty on the relationship between supply chain integration and operational performance. *Journal of Operations Management, 29*(6), 604–615.

Chapter 4
Production Systems and Supply Chain Integration

Abstract The boundary conditions of supply chain integration (SCI) have been widely studied in order to find out when SCI is applicable and effective. Prior studies have paid much attention to external contextual factors, such as supply complexity, environmental uncertainty and country-level infrastructure. This study contributes to the SCI literature by examining the contingency effects of internal production system on the relationship between supplier integration, customer integration and operational performance. Based on organizational information processing theory, we provide evidence to show that the impact of supplier and customer integration on operational performance varies across production systems, including one-of-a-kind production, batch production and mass production systems. The empirical results also reveal how supplier and customer integration can be matched with different configurations of production systems in order to achieve the desired quality, flexibility, delivery or cost performance.

Keywords Supply chain integration · Production system · Operational performance · Boundary condition · Supplier integration · Customer integration

4.1 Introduction

Supply chain integration (SCI) indicates strategic collaboration, information-sharing, joint decision-making and system-coupling between the manufacturer and its supply chain partners, especially in the production phase (Alfalla-Luque et al., 2013; Demeter et al., 2016; Flynn et al., 2010; Kauppi et al., 2016). In prior studies, scholars have confirmed the positive effects of SCI on operational performance (OP). Some studies have examined the contextual conditions under which SCI is effective (Sousa & Voss, 2008). For example, Wong et al. (2011) demonstrate that environmental uncertainty moderates the relationship between SCI and OP. Similarly,

---

This article was published in Journal of Purchasing and Supply Management, Vol. 24, Shou, Y., Li, Y., Park, Y., & Kang, M. Supply chain integration and operational performance: the contingency effects of production systems, pp. 352–360. Copyright Elsevier (2018).

© The Author(s), under exclusive license to Springer Nature Singapore Pte Ltd. 2022   49
Y. Shou et al., *Supply Chain Integration for Sustainable Advantages*,
https://doi.org/10.1007/978-981-16-9332-8_4

Gimenez et al. (2012) indicate that SCI improves performance, but only when the buyer–supplier relationship is characterized by high supply complexity. Wiengarten et al. (2014) extend the literature by considering the role of a country's logistics capability in external SCI.

However, most extant studies on the contextual conditions of SCI focus on the effects of external environmental factors. Few studies have considered internal factors. In a conceptual paper, Ellram et al. (2007) suggest taking product and process characteristics into consideration and establishing an appropriate match between product design, process design and supply chain structure. Tsinopoulos and Mena (2015) assert that manufacturers' internal process structure and product newness require different supply chain configurations at different stages of the product life cycle. Using qualitative data from British manufacturers, Tsinopoulos and Mena (2015) conclude that firms tend to implement supplier integration (SI) and customer integration (CI), especially in the case of high customization and low-volume production, whereas external integration is not particularly necessary for production with high standardization and high volume. In short, it is very important to fit external integration decisions with internal organizational characteristics in order to yield better OP. Nevertheless, limited empirical evidence is available in the extant literature to support the in-depth analysis of how external SCI matches with internal process structures in order to determine individual dimensions of OP.

To address this research gap, this study aims to empirically explore the boundary conditions (Busse et al., 2017) where external SCI (i.e., SI and CI) is effective in terms of quality, flexibility, delivery or cost improvement by analyzing the contingency effects of internal production systems. One-of-a-kind production (OKP), batch production (BP) and mass production (MP) systems are regarded as three main kinds of production system in modern manufacturing practices. Consequently, this study attempts to answer the research question:

RQ How can manufacturing firms utilize external SCI to achieve the desired OP, given their production system configuration?

In response to the research question, this study applies the lens of organizational information processing theory (OIPT). OIPT insists that an organization should align its information-processing capability with information-processing requirements under different conditions (Galbraith, 1973). In prior studies, scholars have investigated SCI as an information-processing system to cope with task interdependence and uncertainty based on OIPT (Hult et al., 2004; Srinivasan & Swink, 2015; Wong et al., 2011). SI and CI can improve an organization's information-processing capability through inter-organizational flows and information-sharing mechanisms (Flynn et al., 2016), as well as by establishing lateral and collaborative relationships with supply chain partners (Srinivasan & Swink, 2015). Production systems, including OKP, BP and MP, differ from each other in terms of the number of variants, lot size, automation, specificity of the equipment, and control of production (Tu & Dean, 2011; Woodward, 1965). We speculate that the operations of different production systems may have distinct information-processing requirements, which should

4.1 Introduction

be matched with different configurations of SI and CI in order to achieve superior OP.

This study contributes to the boundary condition research on SCI. Based on a survey of 791 firms, our findings provide empirical evidence concerning the importance of the fit between SCI and production systems. This study poses both theoretical and managerial implications.

4.2 Theoretical Background and Hypotheses Development

4.2.1 *SCI and Operational Performance*

There is a growing body of research on the theory and practices of SCI. Integration is elaborated as collaboration (which suggests joint goals and collaborative behaviors) and interaction (which indicates formal communication and information exchange). Correspondingly, external SCI is defined as the degree to which a manufacturer strategically collaborates with its suppliers and customers, as well as collaboratively manages cross-firm business processes (Flynn et al., 2010; Wong et al., 2011). In this study, strategic collaboration, information-sharing, joint decision-making and system-coupling are emphasized as key elements of external SCI (Alfalla-Luque et al., 2013; Demeter et al., 2016; Flynn et al., 2010; Kauppi et al., 2016; Wiengarten et al., 2014). Specifically, in an integrative relationship, manufacturers and their supply chain members engage themselves in sharing information about sales forecasts, production plans, order tracking and tracing, delivery status and stock level. Meanwhile, all the supply chain activities are based on risk- or revenue-sharing and long-term agreements. When contingencies happen, they are willing to make decisions jointly and solve problems together in order to maximize benefits for the whole supply chain. In order to achieve the unified control of inter- and intra-firm processes, system-coupling with suppliers and customers, for example, vendor managed inventory (VMI), just-in-time (JIT), kanban and continuous replenishment, is required in the SCI relationship.

According to OIPT, the success of an organization's activities depends on its information-processing capability to deal with uncertain environments (Galbraith, 1973). SI and CI provide the routines and rules with which to coordinate and control the flow of information and to enhance the firm's information-processing capability (Wong et al., 2011) through information-sharing, system-coupling, collaboration and joint decision-making, which can help the firm to perform effectively in terms of cost, quality, delivery and flexibility. In an integrative relationship, supply chain members are willing to exchange and share information about their inventory, production, logistics and sales. Meanwhile, system-coupling among manufacturers, suppliers and customers establishes a platform for the collection and integration of information in the whole supply chain, which in turn improves supply chain transparency and further ensures in-time delivery and supply chain flexibility. In addition, the manufacturer

52 4 Production Systems and Supply Chain Integration

and its suppliers and customers make joint and collaborative decisions on product and process design, quality improvement and cost control when they are involved in an integrative relationship. Information-sharing and system-coupling help the manufacturer to access and integrate information in the supply chain, while joint decision-making enhances manufacturers' capability in processing a wide range of information. Accurate, adequate and timely supply chain information can help reduce inventories, speed up production cycles and improve flexibility (Flynn et al., 2016). Thus, SI and CI contribute to the improvement of OP. Accordingly, we hypothesize that:

H1 SI is positively associated with the firm's OP, including (a) quality, (b) flexibility, (c) delivery and (d) cost.

H2 CI is positively associated with the firm's OP, including (a) quality, (b) flexibility, (c) delivery and (d) cost.

4.2.2 The Contingency Effects of Production System

The production system refers to the arrangement of technological elements (e.g., machines and tools) and organizational behavior (e.g., division of labor and information flow). Based on technical complexity and the continuity of manufacturing, Woodward (1965) classifies production systems into small batch, mass and continuous process production. Later on, given the increasing market demand for customized or individualized products, an OKP system emerged in manufacturing plants (Wortmann, 1992) due to its high flexibility and efficiency. Continuous process production is not typically found in most manufacturing industries. Hence, OKP, BP and MP are considered as three main systems in modern manufacturing firms' production practices. Table 4.1 summarizes the characteristics of the three types of production system.

As shown in Table 4.1, OKP, BP and MP systems differ from each other in terms of flexibility, the number of variants, lot size, automation, specificity of the equipment, and control of production (Hayes & Wheelwright, 1979; Safizadeh et al., 1996; Tu & Dean, 2011; Woodward, 1965). Therefore, we speculate that different configurations of production system indicate different requirements for information.

Specifically, the manufacturer with an OKP-dominated production system delivers highly customized products (e.g., customized windows and doors, structural steel for construction and special industrial equipment) to individual customers (Tu & Dean, 2011). Hence, it needs a large amount of information about product specifications from the customer. CI helps to increase the accuracy of product information, eliminate errors and reworking, and deal with modifications on a timely basis. Therefore, CI contributes to the enhancement of quality, delivery, flexibility and cost-effectiveness in the OKP system. By contrast, SI may not be particularly important to OKP manufacturers in order to achieve the expected OP. The key reason is that OKP manufacturers may have to select suppliers according to specific customer needs. As reported

4.2 Theoretical Background and Hypotheses Development

Table 4.1 Characteristics of production systems

Production system	Product	Lot size	Production conditions	Flexibility
One-of-a-kind production (OKP)	Highly customized products	Small; manufacturing items singly	Relying more on skilled human labor and advanced technology with significant flexibility	High flexibility enables a quick response to customers' individual requirements
Batch production (BP)	Similar products with variants	Middle; manufacturing items by batch	The need to stop and prepare materials, equipment or machines between batches	A certain degree of flexibility should be ensured to produce variants
Mass production (MP)	Highly standard products	Large; manufacturing items by continuous production	Using assembly line techniques and automatic machines to achieve continuous production	A production line with little flexibility is hard to switch, once it is launched

by Tu and Dean (2011), the OKP-dominated production system is characterized by a dynamic supply chain. Under such conditions, as SI can hardly be matched with the information requirements of OKP systems, improvement in OP cannot be guaranteed (Wong et al., 2011).

In a BP-dominated system, products are manufactured in relatively large, but discrete, batches, while the batch size determines the time needed to manufacture the products. Examples of BP systems are the production of food and beverage, the coating, etching and finishing of semi-conductors, and dress-making. Given that a BP system always provides similar products with variants, the manufacturer has to adjust production processes accordingly and manage their procurement and inventory efficiently. Thus, the operations in a BP system require information about orders, materials, inventory, and production capacity from suppliers. In order to cope with complex production processes, manufacturers are encouraged to access the necessary information and improve production capabilities through integration with their suppliers. Meanwhile, information-sharing and system-coupling with suppliers help the manufacturer to make appropriate arrangements for production planning and control in the BP system (Choudhari et al., 2012) and enable the manufacturer to produce goods on time at a low cost (Flynn et al., 2010). Therefore, it can be concluded that SI contributes to OP in a BP-dominated production system. However, unlike the OKP system, products in the BP system are not manufactured for a specific customer but for target customer groups. As a result, the relationship between BP manufacturers and their customers is characterized by a lower level of commitment, which lessens the effectiveness of CI (Zhao et al., 2011).

The MP-dominated system is mainly designed for the continuous production of standardized products, which indicates a low level of uncertainty. In other words, an MP-dominated system indicates low-level information-processing requirements. Information about products and production is relatively certain for MP manufacturers. Therefore, the manufacturer does not urgently need to access real-time information from its suppliers and customers. Consequently, stable and continuous production conditions require low-level information-processing capabilities in the MP-dominated system. OIPT suggests that a simple organizational structure can be efficient in terms of information-processing in a certain environment (Galbraith, 1973). Wong et al. (2011) have also explained how uncertainty moderates the relationship between SCI and OP. Therefore, for the manufacturer with an MP-dominated system, an integrative supply chain structure may not help to improve OP significantly.

To conclude, OKP, BP and MP systems show different needs in terms of information accessing and processing, which may strengthen or weaken the performance effects of external SCI. In other words, the effects of external SCI on OP may be contingent on different production system configurations. Therefore, we propose the following hypotheses:

H3 The relationship between SI and the firm's OP, including (a) quality, (b) flexibility, (c) delivery and (d) cost, varies across production systems.

H4 The relationship between CI and the firm's OP, including (a) quality, (b) flexibility, (c) delivery and (d) cost, varies across production systems.

4.3 Method

4.3.1 Data

We used data collected from the sixth round of the International Manufacturing Strategy Survey (IMSS) to test the proposed hypotheses. Details about the IMSS project are available in Chap. 1 of this book (Shou et al., 2022). In this study, we dropped 140 responses due to missing data in related items, resulting in 791 usable responses.

4.3.2 Measures

All measures for our key constructs were adapted from existing scales in the extant literature. Detailed survey questions are reported in the Appendix of this book (Shou et al., 2022).

Both SI and CI were measured by four items: (1) information-sharing about sales forecast, production plans, delivery status and stock level, (2) collaborative approaches, e.g., risk-/revenue-sharing and long-term agreements, (3) joint decision-making about quality design/modifications, process design/modifications, quality improvement and cost control, and (4) system-coupling of VMI, JIT, kanban and continuous replenishment (Alfalla-Luque et al., 2013; Demeter et al., 2016; Flynn et al., 2010; Kauppi et al., 2016). Five-point Likert scales were used to indicate the current level of SI and CI.

Measures for the four dimensions of OP (i.e., quality, delivery, flexibility and cost) were adapted from Kauppi et al. (2016) and Wiengarten et al. (2014). All the above constructs were measured by five-point Likert scales, in which a higher value indicated better performance. Quality, delivery and cost were measured by two items, while flexibility was measured by three items, since they are all mature constructs. In addition, the IMSS questionnaire included three sections with more than 200 questions. Hence, the length of the questionnaire had to be controlled to ensure that respondents could finish the questions before losing patience.

The measure of production systems was based on a forced-choice instrument in which the plant managers were requested to allocate 100 points across three types of production choices (i.e., OKP, BP and MP). To test the contingency effects of H3 and H4, we divided the sample into three groups, namely, OKP-dominated, BP-dominated and MP-dominated groups, according to the largest values of OKP, BP and MP. There were two equal maximum values in 54 cases (e.g., 40% for OKP, 40% for BP and 10% for MP), which were not dominated by a single production system. After removing these 54 cases, we had 738 responses with which to test H3 and H4. We also ran an analysis with the 54 cases to ensure robustness, and the results indicated that removing these cases did not change our conclusions.

We include three control variables in our study, i.e., firm size, country and industry. Firm size, in terms of the natural logarithm of the total number of employees, is often used as a control variable because larger firms may "take advantages of economies of scale in their business activities" to achieve better performance (Kim & Lee, 2010, p. 964). Country-level characteristics may also influence SCI activities and firm performance (Wiengarten et al., 2014). Lastly, the IMSS data set includes responses from multiple industries. These industries may contribute to the differences in the implementation of SCI and manufacturing firms' OP (Danese & Bortolotti, 2014; Gimenez et al., 2012; Safizadeh et al., 1996; Tsinopoulos & Mena, 2015).

4.3.3 Reliability and Validity

Considering the active involvement of manufacturing firms and the extensive analyses undertaken by a number of important studies based on previous versions of the IMSS survey (Demeter et al., 2016; Flynn et al., 2010; Kauppi et al., 2016; Wiengarten et al., 2014), content validity was established. Exploratory factor analysis (EFA) was executed using SPSS 20.0. The sampling adequacy was confirmed by the value of KMO (0.865), which was above the recommended value of 0.50 (Ferguson & Cox, 1993), while the EFA results indicated that all items had stronger loadings on their respective constructs and lower loadings on other constructs, thus ensuring construct unidimensionality. Cronbach's alpha values were then calculated, ranging from 0.727 to 0.879 (see Table 4.2). The results indicated that the scales were reliable.

To further evaluate convergent and discriminant validity, we used confirmatory factor analysis (CFA) with the maximum likelihood approach, as suggested by O'Leary-Kelly and Vokurka (1998). The CFA results, as summarized in Table 4.2, showed that our proposed structure of the items for measuring the two dimensions of SCI (i.e., SI and CI) and the four dimensions of OP (i.e., quality, flexibility, delivery and cost) resulted in a model with a reasonably good fit, with χ^2/df (3.138) less than 5, with the CFI (0.965), GFI (0.954), IFI (0.966) and TLI (0.955) more than

Table 4.2 Construct reliability and validity analysis

Construct	Item	Mean	SD	Factor loading	Reliability and validity
Supplier integration	SI1	3.312	0.981	0.752	Cronbach's alpha = 0.835; composite reliability = 0.843; AVE = 0.574
	SI2	3.254	1.002	0.826	
	SI3	3.099	1.052	0.777	
	SI4	2.884	1.159	0.668	
Customer integration	CI1	3.143	1.110	0.847	Cronbach's alpha = 0.879; composite reliability = 0.882; AVE = 0.653
	CI2	3.033	1.158	0.855	
	CI3	2.817	1.245	0.777	
	CI4	3.167	1.129	0.747	
Quality	Q1	3.120	0.926	0.838	Cronbach's alpha = 0.842; composite reliability = 0.842; AVE = 0.727
	Q2	3.259	0.967	0.868	
Delivery	D1	3.191	0.969	0.829	Cronbach's alpha = 0.838; composite reliability = 0.840; AVE = 0.724
	D2	3.226	0.991	0.872	
Flexibility	F1	3.240	0.992	0.753	Cronbach's alpha = 0.727; composite reliability = 0.743; AVE = 0.493
	F2	3.161	0.951	0.742	
	F3	3.081	1.006	0.602	
Cost	C1	2.523	0.953	0.814	Cronbach's alpha = 0.742; composite reliability = 0.746; AVE = 0.596
	C2	2.399	0.871	0.727	

4.3 Method 57

0.90, and the RMSEA (0.052) and SRMR (0.035) less than 0.08 (Hu & Bentler, 1999). Furthermore, all factor loadings were greater than 0.50, while the values of composite reliability for all constructs were greater than 0.70. The estimates of the AVE were higher than 0.50 for four constructs and 0.493 for the fifth construct. While the recommended minimum value for the AVE is 0.50 (Fornell & Larcker, 1981), we conclude that our constructs had strong convergent validity based on the above results.

The discriminant validity of the constructs was assessed by comparing the square root of the AVE and the correlation between any pair of them (Fornell & Larcker, 1981). The square root of the AVE of all the constructs is greater than the correlation between any pair of them (see Table 4.3), which provides evidence of discriminant validity. Further, we calculate the heterotrait-monotrait (HTMT) ratio of correlations, as proposed by Henseler et al. (2015), which is "the average of the heterotrait-heteromethod correlations (i.e., the correlations of indicators across constructs measuring different phenomena), relative to the average of the monotrait-heteromethod correlations (i.e., the correlations of indicators within the same construct)" (p. 121). If two constructs are highly correlated with HTMT values close to 1.0, a lack of discriminant validity can be concluded (Henseler et al., 2015; Voorhees et al., 2016). Based on a content analysis of published articles and a Monte

Table 4.3 Correlations of the constructs

Construct	Mean	SD	(1)	(2)	(3)	(4)	(5)	(6)
(1) SI	3.137	0.860	0.758					
(2) CI	3.037	0.995	0.664**	0.808				
(3) Quality	3.192	0.880	0.237**	0.221**	0.853			
(4) Flexibility	3.164	0.791	0.234**	0.240**	0.553**	0.702		
(5) Delivery	3.212	0.909	0.232**	0.234**	0.559**	0.556**	0.851	
(6) Cost	2.462	0.814	0.124**	0.101**	0.271**	0.292**	0.285**	0.771

Note Value on the diagonal is the square root of the average variance extracted (AVE)
**$p < 0.01$

Table 4.4 HTMT results of the correlations

	(1)	(2)	(3)	(4)	(5)	(6)
(1) SI	1					
(2) CI	0.770	1				
(3) Quality	0.275	0.257	1			
(4) Flexibility	0.297	0.299	0.703	1		
(5) Delivery	0.275	0.273	0.667	0.709	1	
(6) Cost	0.156	0.126	0.343	0.395	0.364	1

58   4   Production Systems and Supply Chain Integration

Carlo simulation, Voorhees et al. (2016) suggested a threshold of 0.85 as the criterion in a covariance-based measurement model. Table 4.4 shows the HTMT results, which indicate no discriminant validity problems.

4.4 Results

Our research found both direct effects of SCI on OP and contingency effects of production systems. A structural equation model using AMOS 20.0 was first established to test the SCI-OP relationships (H1 and H2). We then performed a multi-group analysis across OKP-dominated, BP-dominated and MP-dominated groups to examine the contingency effects of production systems (H3 and H4), according to the approach used by Byrne (2013), Nyaga et al. (2010) and Wiengarten et al. (2014). Table 4.5 summarizes the results of the structural model fit for the overall sample and subgroups, with the CFI, and IFI well above the recommended threshold of 0.90, and the RMSEA and SRMR less than 0.80 (Hazen et al., 2015; Wong et al., 2011), indicating a good fit of the data.

The overall sample with 791 cases were used to test H1 and H2, with the results summarized in Table 4.6. All of the four hypotheses of H1 are supported. It can be concluded that SI is positively and significantly associated with product quality, flexibility, delivery and cost. However, the results for H2 are more complicated. The relationship proposed in H2(b) is statistically significant at the level of 0.05, while, in H2(c), it is marginally significant at the level of 0.10. The relationships in H2(a) and H2(d) are not significant in this study. The results indicate that CI is positively associated with production flexibility and delivery, but may not be significantly associated with quality and cost performance.

The contingency effects of production system on the SCI–OP relationship were tested with a comparison of model fits between OKP-dominated, BP-dominated and MP-dominated groups. Following Byrne (2013), we first tested the baseline model (i.e., equal pattern; Model 1 in Table 4.7) with χ^2/df, CFI, TLI and RMSEA being 1.932, 0.927, 0.902 and 0.036, respectively, indicating that the model fits well with the data. In Model 2, each factor loading was forced to be equal across three groups, after which it was compared with Model 1 to test for measurement invariance. The insignificant χ^2 difference between Models 2 and 1 ($\Delta\chi^2 = 27.825$, Δdf $= 22$, $p = 0.182$) shows that factor loadings appeared to be invariant across the three groups.

Table 4.5 Summary of structural model fit indices

Sample		Size	χ^2/df	RMSEA	SRMR	CFI	GFI	NFI	IFI	TLI
Overall sample		791	3.473	0.056	0.068	0.939	0.926	0.917	0.940	0.919
Sub-groups	OKP-dominated	273	2.240	0.068	0.077	0.910	0.876	0.851	0.912	0.879
	BP-dominated	323	1.977	0.055	0.070	0.943	0.901	0.893	0.944	0.924
	MP-dominated	142	1.576	0.070	0.077	0.921	0.842	0.817	0.924	0.894

4.4 Results

Table 4.6 Structural model path coefficients: overall sample

Structural path	Standardized coefficient	p-value	SE	t-value	Result
H1 (a) SI to Quality	0.234***	0.000	0.074	3.326	Supported
H1 (b) SI to Flexibility	0.189**	0.010	0.076	2.590	Supported
H1 (c) SI to Delivery	0.200**	0.005	0.076	2.823	Supported
H1 (d) SI to Cost	0.199**	0.009	0.08	2.622	Supported
H2 (a) CI to Quality	0.094	0.167	0.056	1.383	Not supported
H2 (b) CI to Flexibility	0.161*	0.022	0.058	2.285	Supported
H2 (c) CI to Delivery	0.128†	0.062	0.058	1.866	Marginally supported
H2 (d) CI to Cost	−0.042	0.567	0.061	−0.572	Not supported

$*p < 0.05$, $**p < 0.01$, $***p < 0.001$ and $†p < 0.1$

The invariance of the structural coefficients was tested after the establishment of measurement invariance. Model 3 forced the factor loadings and all path coefficients to be equal across the three groups. The χ^2 difference between Models 3 and 2 was 38.368, with 16 degrees of freedom. The statistically significant result ($p = 0.001$) indicates that structural coefficients across three groups were different, providing support for the contingency effects of production systems (Byrne, 2013).

We then forced specific path coefficients to be equal (i.e., Models 3a-h), and compared each of them with Model 2 in order to identify which path was being impacted by the production system. As shown in Table 4.7, the χ^2 difference between Models 3a and 2 ($\Delta\chi^2 = 7.706$, $\Delta df = 2$, $p = 0.021$) was significant at the level of 0.05, while the χ^2 difference between Models 3c and 2 ($\Delta\chi^2 = 5.592$, $\Delta df = 2$, $p = 0.061$) was significant at the level of 0.10. The results indicate that the paths of SI–quality and SI–delivery were different across the three groups to some extent. The relationship between SI and quality and delivery was only significant in the BP-dominated group. Thus, H3(a) and H3(c) are supported. Moreover, the χ^2 difference between Models 3e and 2 ($\Delta\chi^2 = 11.698$, $\Delta df = 2$, $p = 0.003$) and Models 3f and 2 ($\Delta\chi^2 = 6.057$, $\Delta df = 2$, $p = 0.048$) was statistically significant, while the χ^2 difference between Models 3 g and 2 ($\Delta\chi^2 = 5.132$, $\Delta df = 2$, $p = 0.077$) was marginally significant. The results indicate that the level of association between CI–quality, CI–flexibility and CI–delivery was different across the three groups and only significant for the OKP-dominated group. Therefore, H4(a), H4(b) and H4(c) are supported, revealing the contingency effects of production systems on the paths of CI–quality, CI–flexibility and CI–delivery.

Table 4.7 The contingency effects of production systems using multi-group analysis

Models	χ^2	df	χ^2/df	CFI	TLI	RMSEA	Δdf	$\Delta\chi^2$	Significance level	OKP-dominated (N = 273)	BP-dominated (N = 323)	MP-dominated (N = 142)
1. Equal pattern	1298.077	672	1.932	0.927	0.902	0.036						
2. Equal factor loadings	1325.902	694	1.911	0.926	0.904	0.035	22	27.825	0.182			
3. Equal factor loadings and path coefficients	1364.27	710	1.922	0.924	0.903	0.035	16	38.368	0.001			
3a. SI to Quality	1333.608	696	1.916	0.925	0.904	0.035	2	7.706	0.021	0.062 (0.591, 0.554)	0.506*** (4.734, 0.000)	0.173 (0.715, 0.475)
3b. SI to Flexibility	1326.67	696	1.906	0.926	0.905	0.035	2	0.768	0.681	0.079 (0.750, 0.453)	0.204† (1.900, 0.057)	0.206 (0.844, 0.399)
3c. SI to Delivery	1331.494	696	1.913	0.926	0.904	0.035	2	5.592	0.061	0.040 (0.383, 0.702)	0.378*** (3.656, 0.000)	−0.041 (−0.166, 0.868)
3d. SI to Cost	1328.817	696	1.909	0.926	0.904	0.035	2	2.915	0.233	0.046 (0.414, 0.679)	0.069 (0.639, 0.523)	0.455† (1.912, 0.056)
3e. CI to Quality	1337.600	696	1.922	0.925	0.903	0.035	2	11.698	0.003	0.329** (3.171, 0.002)	−0.185 (−1.872, 0.061)	−0.067 (−0.284, 0.776)
3f. CI to Flexibility	1331.959	696	1.914	0.926	0.904	0.035	2	6.057	0.048	0.362*** (3.405, 0.000)	0.031 (0.297, 0.766)	−0.044 (−0.187, 0.852)
3g. CI to Delivery	1331.034	696	1.912	0.926	0.904	0.035	2	5.132	0.077	0.254** (2.434, 0.015)	−0.067 (−0.687, 0.492)	0.283 (1.176, 0.240)
3h. CI to Cost	1330.415	696	1.912	0.926	0.904	0.035	2	4.514	0.105	0.197† (1.773, 0.076)	−0.132 (−1.254, 0.210)	−0.184 (−0.808, 0.419)

Note: t-values and p-values are in brackets

$*p < 0.05$, $**p < 0.01$, $***p < 0.001$ and $\dagger p < 0.1$

4.4 Results

In conclusion, our results provide strong support for the positive relationships of SI–quality, SI–flexibility, SI–delivery, SI–cost, CI–flexibility and CI–delivery. However, no evidence indicates that CI is positively related to quality performance and cost performance. With regard to the contingency effects, the path coefficients for SI–quality, SI–delivery, CI–quality, CI–flexibility and CI–delivery are significantly different across OKP-dominated, BP-dominated and MP-dominated groups. Table 4.8 summarizes the test results.

4.5 Discussion

4.5.1 Findings

This study explores how external SCI (i.e., SI and CI) influences quality, flexibility, delivery and cost performance, given the different production system configurations. Although prior studies have examined the value of SI and CI in relation to OP, the results of this study provide deeper insights and demonstrate that the effectiveness of external SCI is contingent on the internal production system configuration. Specifically, this study suggests that, for manufacturers with an OKP-dominated system, only CI significantly contributes to quality, flexibility, delivery and cost performance, while SI plays an important role in the improvement of quality, flexibility and delivery performance for BP-dominated manufacturers. Moreover, for manufacturers with an MP-dominated production system, only the SI-cost path coefficient is marginally significant. Overall, the contingency effects of internal production systems are confirmed. Table 4.9 summarizes the findings of this study.

4.5.2 Theoretical Implications

This study contributes to boundary condition research on SCI by investigating the contingency effects of internal production systems on external SCI–OP relationships. As prior studies have reached inconsistent conclusions about the impact of SCI on OP (Flynn et al., 2010), scholars have been encouraged to conduct boundary condition research. To date, scholars have paid a great deal of attention to external contextual factors, such as supply complexity (Gimenez et al., 2012), environmental uncertainty (Wong et al., 2011), and country-level logistics capability (Wiengarten et al., 2014). However, when manufacturing firms make decisions on SI and CI, the organization's internal characteristics, such as product and process structure, should be considered in pursuit of better OP (Ellram et al., 2007; Tsinopoulos & Mena, 2015). Through the establishment of a novel theoretical model and an empirical study based on IMSS data, this study extends the understanding of the boundary conditions of effective SCI from the perspective of internal production systems.

Table 4.8 Summary of hypotheses testing

Structural paths	Overall sample (n = 791)	Direct effects	Subgroups			Moderation effects
			OKP-dominated (n = 273)	BP-dominated (n = 323)	MP-dominated (n = 142)	
	Coefficients		Coefficients	Coefficients	Coefficients	
SI to Quality	0.234***	H1(a) supported	0.062	0.506***	0.173	H3(a) supported
SI to Flexibility	0.189**	H1(b) supported	0.079	0.204†	0.206	H3(b) not supported
SI to Delivery	0.200**	H1(c) supported	0.040	0.378***	−0.041	H3(c) marginally supported
SI to Cost	0.199**	H1(d) supported	0.046	0.069	0.455†	H3(d) not supported
CI to Quality	0.094	H2(a) not supported	0.329**	−0.185	−0.067	H4(a) supported
CI to Flexibility	0.161*	H2(b) supported	0.362***	0.031	−0.044	H4(b) supported
CI to Delivery	0.128†	H2(c) marginally supported	0.254**	−0.067	0.283	H4(c) marginally supported
CI to Cost	−0.042	H2(d) not supported	0.197†	−0.132	−0.184	H4(d) not supported

$*p < 0.05$, $**p < 0.01$, $***p < 0.001$ and $†p < 0.1$

4.5 Discussion

Table 4.9 Summary of effective external integration

	Quality	Flexibility	Delivery	Cost
OKP-dominated system	CI**	CI***	CI**	CI†
BP-dominated system	SI***	SI†	SI***	–
MP-dominated system	–	–	–	SI†

*$p < 0.05$, **$p < 0.01$, ***$p < 0.001$ and †$p < 0.1$

The findings of this study reveal how SI or CI is matched with different production system configurations in order to achieve the desired quality, flexibility, delivery or cost performance. Although a few studies have advocated a match between process structure and supply chain configuration (Ellram et al., 2007; Tsinopoulos & Mena, 2015), our research presents empirical evidence for the differential effects of SI and CI on the four dimensions of OP, given a specific production system configuration. Specifically, CI for an OKP-dominated manufacturer helps to enhance overall OP. Thus, this study contributes to the knowledge about how OKP manufacturers simultaneously maintain high customization and high OP. Additionally, SI for a BP-dominated manufacturer helps to increase quality, flexibility and delivery performance, while, in an MP system, SI may help to improve cost performance. The findings improve our understanding of the relationship between production systems and SCI.

Moreover, the findings extend the application of OIPT when studying supply chain management phenomena. SCI has been confirmed as an information-processing system in response to task interdependency and uncertainty in the extant literature (Bode et al., 2011; Fan et al., 2017; Hult et al., 2004; Srinivasan & Swink, 2015). The effectiveness of SCI relies on its proper and rational alignment with information-processing requirements of the organization under the contextual conditions of task complexity and task security (Narayanan et al., 2011), product and market complexity (Wong et al., 2015), environmental uncertainty (Wong et al., 2011), and supply chain uncertainty (Flynn et al., 2016). However, to the best of our knowledge, there is a limited body of literature on the information-processing requirements of production systems and the fit between SCI and production systems from the OIPT perspective. This study argues that OKP, BP and MP systems differ in terms of production processes, technical complexity and the continuity of manufacturing, such that they have different requirements concerning demand and supply information. Our results indicate that OKP with CI and BP with SI make the best contribution to OP. The findings provide a nuanced understanding of the performance impact of SI and CI when they are coupled with the information-processing needs of a specific production system configuration. The empirical findings on the fit between external SCI (SI and CI) and internal production systems (OKP, BP and MP) further support the theoretical logic of OIPT, while complementing prior studies (e.g., Bode et al., 2011; Flynn et al., 2016; Narayanan et al., 2011; Srinivasan & Swink, 2015; Wong et al., 2011, 2015).

4.6 Conclusions

In summary, this study shows how the fit between external SCI and internal production systems determines individual dimensions of OP. The findings contribute to the boundary condition research on SCI by focusing on the contingency effects of production systems. In addition, this study provides a nuanced understanding of the fit between external SCI and OKP, BP and MP systems from the perspective of OIPT, which in turn extends the application of OIPT to supply chain management research. This study also provides some managerial implications. Manufacturers are recommended to make integration decisions based on their internal production system configurations in order to achieve the desired quality, flexibility, delivery or cost performance.

This study has some limitations, which suggest opportunities for future research. First, while previous studies have focused on external environmental contexts of SCI, this study emphasizes the role of the internal production system. However, manufacturers' decisions on SCI choices may rely on both external and internal contingency factors. Thus, it would be interesting to explore how both external and internal factors simultaneously influence the effectiveness of SCI. Second, model testing in this study is based on a cross-sectional design, with data exclusively collected from manufacturers. Since SCI generally requires time to develop, it is difficult to examine the dynamic characteristics of SCI by using cross-sectional data. Thus, further longitudinal studies could contribute to the development of useful insights into the evolution and long-term effects of SCI. In addition, data collected from suppliers and customers could offer more comprehensive knowledge. Third, besides OP, sustainability has been a major concern among supply chain executives. Future research is suggested to explore how the match between SCI and production systems affects sustainability.

Acknowledgements This work was supported by the National Natural Science Foundation of China under Grant number 71472166.

References

Alfalla-Luque, R., Medina-Lopez, C., & Dey, P. K. (2013). Supply chain integration framework using literature review. *Production Planning & Control, 24*(8–9), 800–817.

Bode, C., Wagner, S. M., Petersen, K. J., & Ellram, L. M. (2011). Understanding responses to supply chain disruptions: Insights from information processing and resource dependence perspectives. *Academy of Management Journal, 54*(4), 833–856.

Busse, C., Kach, A., & Wagner, S. M. (2017). Boundary conditions: What they are, how to explore them, why we need them, and when to consider them. *Organizational Research Methods, 20*(4), 574–609.

Byrne, B. M. (2013). *Structural equation modeling with AMOS: Basic concepts, applications, and programming*. Routledge.

References

Choudhari, S. C., Adil, G. K., & Ananthakumar, U. (2012). Choices in manufacturing strategy decision areas in batch production system—Six case studies. *International Journal of Production Research, 50*(14), 3698–3717.

Danese, P., & Bortolotti, T. (2014). Supply chain integration patterns and operational performance: A plant-level survey-based analysis. *International Journal of Production Research, 52*(23), 7062–7083.

Demeter, K., Szász, L., & Rácz, B. G. (2016). The impact of subsidiaries' internal and external integration on operational performance. *International Journal of Production Economics, 182*, 73–85.

Ellram, L. M., Tate, W. L., & Carter, C. R. (2007). Product-process-supply chain: An integrative approach to three-dimensional concurrent engineering. *International Journal of Physical Distribution & Logistics Management, 37*(4), 305–330.

Fan, H., Li, G., Sun, H., & Cheng, T. C. E. (2017). An information processing perspective on supply chain risk management: Antecedents, mechanism, and consequences. *International Journal of Production Economics, 185*, 63–75.

Ferguson, E., & Cox, T. (1993). Exploratory factor analysis: A users' guide. *International Journal of Selection and Assessment, 1*(2), 84–94.

Flynn, B. B., Huo, B., & Zhao, X. (2010). The impact of supply chain integration on performance: A contingency and configuration approach. *Journal of Operations Management, 28*(1), 58–71.

Flynn, B. B., Koufteros, X., & Lu, G. (2016). On theory in supply chain uncertainty and its implications for supply chain integration. *Journal of Supply Chain Management, 52*(3), 3–27.

Fornell, C., & Larcker, D. F. (1981). Evaluating structural equation models with unobservable variables and measurement error. *Journal of MarketIng Research, 18*(1), 39–50.

Galbraith, J. R. (1973). *Designing complex organizations.* Addison-Wesley.

Gimenez, C., van der Vaart, T., & Pieter Van Donk, D. (2012). Supply chain integration and performance: The moderating effect of supply complexity. *International Journal of Operations & Production Management, 32*(5), 583–610.

Hayes, R. H., & Wheelwright, S. C. (1979). Link manufacturing process and product life cycles. *Harvard Business Review, 57*, 133–140.

Hazen, B. T., Overstreet, R. E., & Boone, C. A. (2015). Suggested reporting guidelines for structural equation modeling in supply chain management research. *International Journal of Logistics Management, 26*(3), 627–641.

Henseler, J., Ringle, C. M., & Sarstedt, M. (2015). A new criterion for assessing discriminant validity in variance-based structural equation modeling. *Journal of the Academy of Marketing Science, 43*(1), 115–135.

Hu, L., & Bentler, P. M. (1999). Cutoff criteria for fit indexes in covariance structure analysis: Conventional criteria versus new alternatives. *Structural Equation Modeling, 6*, 1–55.

Hult, G. T. M., Ketchen, D. J., & Slater, S. F. (2004). Information processing, knowledge development, and strategic supply chain performance. *Academy of Management Journal, 47*(2), 241–253.

Kauppi, K., Longoni, A., Caniato, F., Kuula, M., & Grubbström, R. W. (2016). Managing country disruption risks and improving operational performance: Risk management along integrated supply chains. *International Journal of Production Economics, 182*, 484–495.

Kim, D., & Lee, R. P. (2010). Systems collaboration and strategic collaboration: Their impacts on supply chain responsiveness and market performance. *Decision Sciences, 41*(4), 955–981.

Narayanan, S., Jayaraman, V., Luo, Y., & Swaminathan, J. M. (2011). The antecedents of process integration in business process outsourcing and its effect on firm performance. *Journal of Operations Management, 29*(1), 3–16.

Nyaga, G. N., Whipple, J. M., & Lynch, D. F. (2010). Examining supply chain relationships: Do buyer and supplier perspectives on collaborative relationships differ? *Journal of Operations Management, 28*(2), 101–114.

O'Leary-Kelly, S. W., & Vokurka, R. J. (1998). The empirical assessment of construct validity. *Journal of Operations Management, 16*(4), 387–405.

Safizadeh, M. H., Sharma, D., Sharma, D., & Wood, C. (1996). An empirical analysis of the product-process matrix. *Management Science, 42*, 1576–1591.

Shou, Y., Kang, M., & Park, Y. (2022). *Supply chain integration for sustainable advantages.* Springer.

Sousa, R., & Voss, C. A. (2008). Contingency research in operations management practices. *Journal of Operations Management, 26*(6), 697–713.

Srinivasan, R., & Swink, M. (2015). Leveraging supply chain integration through planning comprehensiveness: An organizational information processing theory perspective. *Decision Sciences, 46*(5), 823–861.

Tsinopoulos, C., & Mena, C. (2015). Supply chain integration configurations: Process structure and product newness. *International Journal of Operations and Production Management, 35*(10), 1437–1459.

Tu, Y., & Dean, P. (2011). *One-of-a-kind production.* Springer.

Voorhees, C. M., Brady, M. K., Calantone, R., & Ramirez, E. (2016). Discriminant validity testing in marketing: An analysis, causes for concern, and proposed remedies. *Journal of the Academy of Marketing Science, 44*(1), 1–16.

Wiengarten, F., Pagell, M., Ahmed, M. U., & Gimenez, C. (2014). Do a country's logistical capabilities moderate the external integration performance relationship? *Journal of Operations Management, 32*(1), 51–63.

Wong, C. Y., Boon-Itt, S., & Wong, C. W. (2011). The contingency effects of environmental uncertainty on the relationship between supply chain integration and operational performance. *Journal of Operations Management, 29*(6), 604–615.

Wong, C. W. Y., Lai, K., & Bernroider, E. W. N., 2015. The performance of contingencies of supply chain information integration: The roles of product and market complexity. *International Journal of Production Economics, 165*(C): 1–11.

Woodward, J. (1965). *Industrial organization: Theory and practice.* Oxford University Press.

Wortmann, J. C. (1992). Production management systems for one-of-a-kind products. *Computers in Industry, 19*(1), 79–88.

Zhao, X., Huo, B., Selen, W., & Yeung, J. H. Y. (2011). The impact of internal integration and relationship commitment on external integration. *Journal of Operations Management, 29*(1), 17–32.

Chapter 5
Enablers of Supply Chain Integration: A Socio-Technical System Perspective

Abstract Although supply chain integration (SCI) and its antecedents has been extensively studied, whether digital manufacturing technology (DMT) will support SCI in this ever increasingly competitive era and how this link may work is still not clear. Drawing on the socio-technical system (STS) perspective, we propose that human resource (HR) is an important social counterpart of DMT. Besides, this study also introduces industrial competition as an environmental factor that may influence the way of how social and technical factors will support SCI. By adding the infrequently discussed environmental part into this study on SCI implementation, a more comprehensive model is proposed and tested with data from the International Manufacturing Strategy Survey (IMSS) project database. The results show that DMT and HR have significant and positive effects on the three dimensions of SCI. Competition moderates the effects of HR on SCI dimensions differentially but do not influence the relationship between DMT and SCI. By investigating the effects of HR and DMT on SCI in the context of competition, this study contributes to the relevant literature.

Keywords Supply chain integration · Digital manufacturing technology · Human resource · Competition · Socio-technical system

5.1 Introduction

The fierce global competition among various industries is becoming a haunting problem of firms all over the world. Manufacturing firms are under increasing pressure for they should outperform their rivals by not only excellent internal operations but also competitive supply chains. Sharing information and coordinating supply chain partners may bring various benefits in terms of cost, flexibility, resilience and so on (Prajogo & Olhager, 2012; Van Donk et al., 2017). Therefore, an increasing number of top managers treat the comprehensive managing practice, supply chain integration (SCI), as a silver bullet and it has become a hot topic of operations management academia in recent decades (Alfalla-Luque et al., 2013). However,

---

This chapter is a revised version of the following journal paper: Tian, M., Huo, B., Park, Y., & Kang, M. (2021). Enablers of supply chain integration: a technology–organization–environment view. *Industrial Management & Data Systems, 121*(8), 1871–1895.

© The Author(s), under exclusive license to Springer Nature Singapore Pte Ltd. 2022
Y. Shou et al., *Supply Chain Integration for Sustainable Advantages*,
https://doi.org/10.1007/978-981-16-9332-8_5

the successful implementation of SCI has always been a tough job (Teller et al., 2012). Thus, many enablers of SCI, in terms of product, resources, environmental and combined dimensions, have been identified in the extant literature (e.g., Shou et al., 2017; Xu et al., 2014; Zhang et al., 2019).

The socio-technical system (STS) perspective highlights the positive effects of simultaneous focus on social and technological factors, both of which have been widely examined in the SCI literature. For example, Xu et al. (2014) identified top management support and information technology as two critical enablers. Huo et al. (2016a) suggested strategic supply chain relationships and supply chain technology are important social and technical factors to foster information integration. However, with the diffusion of emerging technologies and the new business environments they bring, we need to update our understanding of the possible drivers from diversified and complex contexts with the STS perspective (Closs et al., 2008). Although the STS perspective originates with its open system idea and strengthens the impact of surrounding environment (Pasmore et al., 1982), existing studies show scarce inclusion of the factors from the environmental context (Kull et al., 2013).

In this study, we include digital manufacturing technology (DMT) as the representative of technical factors. More and more manufacturers have used DMT to pursue a higher level of efficiency, flexibility, and cost-effectiveness, and achieve connectivity, automation, and computation (Dalenogare et al., 2018). DMT may reshape the interactive processes in the job shops and establish new routines (Aversa et al., 2021). Typical digital technologies include real-time tracking and data transmission by radio-frequency identification (RFID), three-dimensional (3D) printing, precision technologies, adaptive manufacturing systems, automated processes, and robotics (Dalenogare et al., 2018; Strozzi et al., 2017). There are different stages and various patterns about how to implement DMT, which makes the manufacturing processes become more complicated and more intelligent (Frank et al., 2019). Researchers have argued that digitalization of manufacturing processes and DMT could integrate various information of product development, delivery, and even customer usage (Holmström et al., 2019). However, whether a firm can benefit from keeping up with the emerging technologies like DMT remains uncovered.

There are collaborative uses of DMT to improve both operations management and supply chain efficiency in the manufacturing industry. For example, the smart factory started by Philips is building demand-driven supply chains by forming an effective collaborating network with its partners, and Nokia's conscious factory makes use of worldwide real-time information to precisely predict maintenance and "collaborate seamlessly with external parties in areas like order tracking" (Geissbauer et al., 2017, p. 14). Companies like Fujitsu provide some ongoing progress about the "digital integration of supply chains with their engineering chains" (Fujitsu Components America, 2021). By implementing DMT, firms could achieve smart manufacturing with a wider reach beyond the factory and gain a clear competitive advantage with faster new product development (like digital creation, development and execution of new products), better customer service, quicker resilience and more cost reduction. The applications of DMT might create vital digital connections with stakeholders in the wider value chain and change shop floors by either being "highly integrated into partners' broader supply chains and digital transformation agendas, or focus(ing) on 'batch size one' manufacturing by providing digital configuration of products through specialist online platforms to achieve market access" (Fujitsu Components America, 2021).

Besides, we focus on human resource (HR) as the key social factor, considering the significance of people in implementing SCI (Huo et al., 2015a). In addition, firms inevitably face increasingly intense competition nowadays. Facing a highly competitive environment, firms' resources are limited and firms should allocate their resources to key processes or activities as well as identify the most effective approach to achieve SCI. Thus, it is meaningful to involve competition, as an environmental factor, to investigate its moderating role in the relationship between HR/DMT and the three dimensions of SCI (i.e., internal, supplier and customer integration).

Based on the guidance of the STS perspective, this study aims to answer two important research questions (RQs) about the successful implementation of SCI:

RQ1 Will HR and DMT (i.e., the social and technical factors) improve the three dimensions of SCI directly?

RQ2 What role does the competition play in the relationship between HR/DMT and the three SCI dimensions?

With the data from the International Manufacturing Strategy Survey (IMSS) project, we could obtain empirical evidence and reply to the above-mentioned research questions. There are several contributions provided by this study: First, this study adds more insights to the SCI literature by enhancing the understanding of improving SCI practices in manufacturing firms with the social and technical antecedents and illustrating the exact moderating effects of the competitive environment, which is faced by more and more manufacturing firms. Next, this research contributes to the STS perspective by considering not only the direct enablers but also the impact of an important environmental factor, competition. Lastly, this study enriches the discussion about DMT by demonstrating the important positive role of these emerging technologies on promoting strategic activities like SCI.

This chapter is structured as follows. Section 5.2 first reviews the literature about STS theory, HR and DMT by introducing how existing studies discuss their definitions, core ideas, antecedents and/or influences. In this way, research gaps are identified and then the hypotheses and conceptual model are developed based on the STS perspective. To be more specific, this study attempts to test the coexisting technical and social support of SCI and discuss the influence of competing environment on these relationships. Section 5.3 elaborates the way of data collection and analysis with a rigorous method. Results are presented in Sect. 5.4. Section 5.5 presents the discussion including theoretical contributions. In the end, limitations of this study and suggestions for future research are provided in Sect. 5.6.

5.2 Literature Review and Hypotheses Development

5.2.1 Socio-Technical System Perspective

In order to develop successful projects or practices in organizations, managers are suggested to take the STS perspective to understand how new systems work in certain environments (Bostrom & Heinen, 1977). A general organization comprises two main subsystems, the social one and the technical one, both of which are enveloped by the environmental subsystem (Kull et al., 2013; Pasmore et al., 1982). The social and technical subsystems support organizational processes simultaneously. For instance, Chaudhuri and Jayaram (2019) studied the social and technical integration, both of which contribute to quality and sustainability programme deployment and lead to better performance. Huo et al. (2015b) found that relationship commitment and information technologies between focal firms and supply chain members enhance supply chain coordination and then supply chain performance. When organizations operate smoothly, the social part might be overlooked as normal routines and the imbalanced importance of technologies are realized. On the other hand, some organizations focus exclusively on the social side and treat technologies as constant when they pursue effectiveness (Pasmore et al., 1982). However, for firms who want to achieve competitive advantages by managing projects or practices excellently, careful consideration about technical means, social interactions and environmental characters is imperative.

Subsystems of an STS contain abundant sets of ideas (Shou et al., 2021). To be more specific, the technical subsystem encompasses all tools, devices, procedures, methods, techniques, and knowledge that involves in value-adding operations (Chaudhuri & Jayaram, 2019; Kull et al., 2013). The social subsystems refers to actors, teams, structures and their behaviors, attitudes, relations, cultures and etc. (Shou et al., 2021; Siawsh et al., 2021). The STS perspective strengthens that there should be open systems (Pasmore et al., 1982), which means both social and technical subsystem are nested in the relevant environment like governmental, societal, industrial and economic ones (Kull et al., 2013).

The STS perspective states that different results emerge from interrelation between subsystems (Pasmore et al., 1982). Joint optimization, by which the social and technical subsystem can build reciprocal interactions and generate optimal outcomes for organizations, is one of the underlined requirements and also a center goal of the STS (Emery, 1959). Manz and Stewart (1997) proposed that social subsystems can be complementary to total quality management (TQM) techniques to create flexible stability inside organizations and finally bring excellent performance. Shou et al. (2021) took a nuanced look at the STS of traceability and supply chain coordination and their results indicated a superior performance when two subsystems match each other.

The STS is regarded as a useful framework to understand how organizations implement good SCI (Chen et al., 2009). There are bundles of studies that take one or more than one sides of the STS to figure out the useful enablers or drivers of

5.2 Literature Review and Hypotheses Development

SCI. In brief, there are four streams of research from different STS perspectives. For the social dimension, both individual and organizational level resources will influence the implementation of SCI. For example, employee commitment (Alfalla-Luque et al., 2015), employee satisfaction (Jacobs et al., 2016), employee participation (Huo et al., 2015a) and employee or manager skills (Huo et al., 2016b) have direct benefits on SCI. Meanwhile, internal communication climate (Zsidisin et al., 2015) and organizational commitment (Huo et al., 2016b) are also regarded as important organizational level factors. As for the technical dimension, information technology is most frequently investigated for its positive effects on SCI and integration with other functional departments or service providers (Liu et al., 2015; Narayanan et al., 2011; Prajogo & Olhager, 2012). Other technical supports consist of advanced manufacturing technologies (Moyano-Fuentes et al., 2016) and functional applications (Ganbold et al., 2020). Environmental factors, like uncertanty (Bae, 2017) and supply chain risks (Jajja et al., 2018; Zhao et al., 2013) are also notable in the SCI literature.

Another stream is from a combined viewpoint, which argues that the co-existance of STS subsystems can have influence on SCI. Xu et al. (2014) suggested and verified that top management support and information technology will boost both supplier and customer integration. Another study investigated whether the STS works in the supply chain context and their results delineate the positive relationship both from strategic supply chain relationship and supply chain technology internalization to supply chain information integration (Huo et al., 2016a), which sent a similar message with Wang et al. (2016). In addition, Birasnav and Bienstock (2019) showed the significant relationship between both leadership or advanced manufacturing technologies and SCI. Besides, Kull et al. (2013) proposed that supplier integration means the "merging of technical systems across firms" and inappropriate design will lead to behavior constraints rather than better integration under some environmental contingencies. Although many studies considered the SCI problem from the STS perspective, how the social and technical antecedents interact with environmental factors is not yet fully studied with empirical data. Therefore, this study follows the guideline of the STS theory and expects to discuss how firms could realize better SCI according to their social, technical and environment characters' interactions.

5.2.2 Human Resources

Wright et al. (1994) defined HR as the pool of human capital under a firm's control. They identified two aspects of HR: employee skills, including each individual's knowledge, skills and abilities, and employee behavior, through which employee skills provide value to the organization. Therefore, HR is a comprehensive reflection of organizations' social subsystem and has a fundamental effect on the effective usage of almost every practice, process and technology. According to MacDuffie (1995), investing in HR yields better performance because employees with well-developed skills are motivated to apply their knowledge and abilities in appropriate contexts. It is also necessary to consider the role of employee incentives. According to Batt (2002), a high-involvement HR system contains three dimensions: (1) employee skills that can be used to complete specific tasks, (2) work design that provides opportunities for individual ongoing learning and improvement, and (3) HR incentives, such as investment in training or high relative pay. These dimensions can be referred to as employee skills, employee participation and employee incentives.

Studies have examined the roles of different aspects of HR in SCI, such as top management support (Xu et al., 2014), high-involvement human resource management (HRM) practices including employee skills, incentives and participation (Huo et al., 2015a), and human capital, including organizational commitment and multi-skilling (Huo et al., 2016b). As a source of competitive advantages (Wright et al., 1994), HR plays a key role in building and organizing relationships within supply chains (Lengnick-Hall et al., 2013). They are also strategic resources that should produce beneficial outcomes in supply chains (Ellinger & Ellinger, 2014).

5.2.3 Digital Manufacturing Technology

DMT refer to enabling technologies (i.e., Internet of Things, cloud computing, cyber-physical systems, and blockchain) that are necessary for the realization of digital manufacturing (Liao et al., 2017). Digital manufacturing, also known as smart manufacturing in Industry 4.0, refers to that manufacturing firms generate, store, analyze and use data of production operations, (i.e., design, production, sourcing, inventory, delivering, maintenance and etc.) to achieve high efficiency, flexibility and customization (Agrifoglio et al., 2017; Lorenz et al., 2020; Xu et al., 2018). DMT is the emerging technical part of the whole technical subsystem of organizations and has enormous impact of the way manufacturers operate and produce. The application of DMT can build novel combination between focal firms and supply chain partners, logistics service providers and end users (Holmström & Partanen, 2014).

Studies on DMT have evolved from the descriptive studies including the overview of DMT (Buer et al., 2021; Strozzi et al., 2017), the suggestion for future direction in the domain of DMT (Hofmann & Rüsch, 2017), and general concepts of digital manufacturing (Yin et al., 2018) to empirical studies focusing on the antecedents or consequences of DMT. For example, Gillani et al. (2020) identified the drivers from technological context, organizational context, and environmental context which boost the applications of DMT as well as the effective role of DMT in improving the firm's operational performance. Kamble et al. (2018) identified the main barriers of DMT adoption, including lack of employee skills, high implementation cost, resistance of organizational and process changes, and lack of strategy. Researchers also pointed out that large firms are the main adopters of DMT (Buer et al., 2021). Bokrantz et al. (2017) identified different driving factors that promote the realization of digital manufacturing, including the advancement of data analytics, increased emphasis on education and training, and stronger environmental legislation and standards. Dalenogare et al. (2018) provided empirical support that DMT can bring

5.2 Literature Review and Hypotheses Development

in benefits in terms of product, operational, and environmental and social side-effects for organizations. Szalavetz (2019) also suggested that DMT can enhance the firm's production capability. In sum, DMT can bring in several benefits to the overall capabilities or performances of organizations. However, whether the manufacturer's input in DMT can help the firm promote specific strategic activities or building strategic capability and if this relationship will be strengthened or weakened by some contextual factors are open questions that need further investigation.

5.2.4 Effects of HR on SCI

HR could be an important driver of all the three dimensions of SCI (i.e., internal integration, supplier integration and customer integration). Internal integration requires the involvement of employees from different functions and positions. Training programs may be necessary to help employees understand the significance of SCI (Ellinger & Ellinger, 2014). Employees with high flexibility, such as multitasking, multi-skilling, and job rotation, can help firms to better deal with SCI problems (Menon, 2012; Pagell, 2004). Training and learning throughout the company provide opportunities to exchange information and ideas, which helps to cultivate good relationships between employees (Wang et al., 2016). Apart from the direct improvement of employees, team or structural level HR efforts are also important to the successful implementation of internal integration. For instance, having a goal-based incentive system that aligns self-interests with organizational objectives can make a significant difference in integrating activities (Pagell, 2004). Likewise, building autonomous teams that include people from different functions can facilitate knowledge transfer and information sharing between functions (Cohen & Levinthal, 1990). Hitt et al. (1993) also suggested that cross-functional appointments/training and multifunctional teams promoted inter-functional integration. Therefore, we propose the following hypothesis:

H1a HR is positively related to internal integration.

External integration focuses on building long-term relationships with suppliers and customers through coordinating strategies and actions, and sharing information (Zhao et al., 2011). In establishing such strategic relationships with supply chain partners, firms could cultivate a deep understanding of their partners (McAfee et al., 2002). Ellinger et al. (2010) argued that training programs help employees better understand the operational activities of their suppliers and customers and develop a common background of knowledge. With such training, employees can help supply chain partners identify problems, make effective decisions and facilitate the dissemination and assimilation of information across the supply chain, resulting in suppliers' and customers' effective engagement in external integration (Cohen & Levinthal, 1990; Huo et al., 2015a; Menon, 2012). Furthermore, employees in autonomous teams influenced by cooperative culture are likely to include suppliers and customers in their groups, thereby facilitating inter-firm communication and collaboration (Waldman, 1994). Although external integration in terms of different partners may have diverse foci on the activities, the benefits from HR efforts are valuable. Therefore, we propose:

H1b HR is positively related to supplier integration.

H1c HR is positively related to customer integration.

5.2.5 Effects of DMT on SCI

DMT is an umbrella term that includes various enabling technologies that use computers and other computing devices like cloud platforms to control, track, monitor, evaluate, and guide operating activities, either directly or indirectly. DMT enhances internal integration in three ways. First, DMT is used to automate design, fabrication, assembly, and material handling processes (Vinodh et al., 2009), which relieves employees from repeated and tedious work to creative ones and promotes workplace communication and information sharing. Second, DMT with synchronized processes and standardized interfaces can automatically collect data from the digital manufacturing equipment and visualize these data to decision-makers, which significantly strengthens the transparent and seamless exchange of information within the firm and thereby allow different functional departments to coordinate more effectively (Boyer et al., 1997; Zhou et al., 2009). Third, the development of a digital factory, which is defined as "an IT system capable of digitally planning, controlling and optimizing all resources and activities related to a product" (Himmler, 2014, p. 17), helps people from different functions quickly understand the manufacturing processes so that they can form project-oriented teams and work together efficiently (Moyano-Fuentes et al., 2016). Therefore, we propose:

H2a DMT is positively related to internal integration.

DMT is widely applied in manufacturing processes that have automated and well-designed control. It helps to ensure information processing flows and improves the quality of data, including inventory, production, and delivery information (Wang et al., 2016). Having accurate and effective information and sharing it with supply chain members enable suppliers to respond quickly and flexibly to manufacturers' needs, thereby resulting in improved coordination. Using computer-aided technology benefits product design and manufacturing by building a platform (Liker et al., 1998), which improves the suppliers' early involvement in new product development projects. Additionally, DMT applied in the manufacturing phase can help ensure product quality, which helps to build and maintain good relationships with customers (Boyer et al., 1997). Transparent manufacturing processes based on DMT also help reduce the conflict between supply chain members (Sodhi & Tang, 2019). Furthermore, DMT integrated into unified systems is credited with responding effectively to changing customer needs, which promotes customer integration (Kathuria, 2000). Therefore, we propose:

5.2 Literature Review and Hypotheses Development 75

H2b DMT is positively related to supplier integration.

H2c DMT is positively related to customer integration.

5.2.6 The Moderating Effect of Competition

The manufacturer operates as a part of its environment, and its decision is shaped by its specific context (DiMaggio & Powell, 1983). Following Oliveira et al. (2019), we propose that the environmental context can affect organizations and their subsystems in terms of decision-making, resource allocating, cross-functional interactions and daily operations. Competition, as one of the manufacturer's significant cognitions of its environmental characters, can influence the manufacturer's decision-making significantly (Venkatesh & Bala, 2012). Therefore, this study focuses on how external competition influences the social and technical subsystems (i.e., HR and DMT).

First, we argue that competition, as a kind of environmental factor, plays a positive moderating role in the relationship between social or technical factors and internal integration. The success of implementing a practice depends on the needs in the firm as well as demands from the environment. A high level of competition indicates that manufacturers may encounter frequent price wars and the risk of product substitution and being cut off by major customers (Gillani et al., 2020). To better deal with those uncertainties, the manufacturer must consolidate the internal functions to maintain profitability and respond swiftly (Zhao et al., 2011). Therefore, when competition is high, the manufacturer will form a high motivation for internal integration. Influenced by this motivation, employees from different functions are more likely to realize the importance of internal integration by making a greater effort in achieving frequent information exchange and communicating among internal functions. DMT will be better utilized to implement internal integration in order to improve efficiency. In sum, a high level of competition can push the organizations to realize the urgency and significance of internal integration. Under this circumstance, employees are encouraged to update useful knowledge and refine their skills, and the firm is stimulated to make better use of DMT to achieve higher internal integration. Thus, both HR and DMT can play a greater role in promoting internal integration. Therefore, we propose that:

H3a Competition positively moderates the effect of HR on internal integration.

H4a Competition positively moderates the effect of DMT on internal integration.

We also argue that competition can influence the roles of HR and DMT in promoting supplier and customer integration. Focal firms are eager to build solid external integration to ensure that all partners follow their strategic or tactical changes closely. The fierce competition requires flexible supply chains, which also need the high level of external integration in turn. In addition, one of the critical components of the competition perceived by the manufacturer is the bargaining power of its major suppliers and customers. A high level of competition indicates that the manufacturer may face a high level of bargaining power from its major partners and its daily operations will also be largely intervened by partners' coercive tactics (Handley & Benton, 2012). In this situation, the manufacturer is less independent and greatly relies on the relationship with major suppliers and customers. To mitigate the concern from business partners, the manufacturer will take actions to build good partnerships and integrate with them (Wang et al., 2021). High level external integration can form close cooperation and bring deep mutual understanding between supply chain partners, which is of great importance when firms are under fierce competition. Realizing the critical significance of external integration, employees will more proactively engage in interactive activities, such as information sharing and communication. A strong motivation for integration with partners will further help the employees overcome the difficulties of integration.

Besides, in a highly competitive environment, it becomes more difficult for manufacturers to survive and succeed. To win from competition, firms are required to continuously improve productivity. As competitors increase, the firm's access to resources is shrinking, resulting in a resource restriction dilemma for the firm. Typically, firms are faced with cost pressure and resource restriction when competition is high. Considering that resources are limited, firms will design a scientific way of resource utilization to achieve the best use of their resources. Especially, DMT has the potential of effective resource deployment and further optimization of the firm's production processes with supply chain partners. When competition is severe, the manufacture has a strong desire to make better use of DMT. Therefore, we propose that:

H3b Competition positively moderates the effect of HR on supplier integration.

H4b Competition positively moderates the effect of DMT on supplier integration.

H3c Competition positively moderates the effect of HR on customer integration.

H4c Competition positively moderates the effect of DMT on customer integration.

Figure 5.1 depicts the conceptual model.

5.3 Method

5.3.1 Data

To test the proposed hypotheses, we used data from the sixth-round IMSS. Please refer to Chapter 1 of this book (Shou et al., 2022) for more information about this dataset, which was developed with a rigorous process. A total of 931 responses were obtained from the IMSS dataset and used in this study after dropping the responses with more than 60% missing value.

5.3 Method

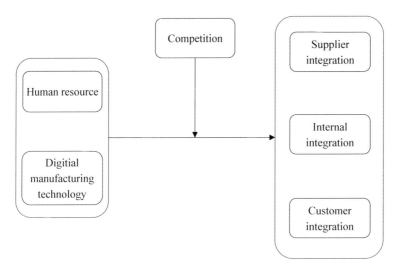

Fig. 5.1 Conceptual model

5.3.2 Measures

We followed previous studies to measure our constructs. Detailed survey questions are available in the Appendix of this book (Shou et al., 2022).

The three dimensions of SCI were measured by asking for the current level of implementation of relative activities such as information sharing, process coordination, and joint decision-making (Shou et al., 2017). HR was measured by assessing the level of implementation of HRM practices. DMT was measured by indicating the current level of implementation of advanced equipment, production processes, and process control in manufacturing. Finally, following Gillani et al. (2020), we operationalized competition as a composite indicator, which depicted the intensity of competitive rivalry, the difficulty of market entry, the bargaining power of customers and suppliers, and the threat of product substitution. Firm size is included as a control variable, which was measured by the natural logarithm of the number of employees.

Table 5.1 shows the descriptive statistics of all variables.

5.3.3 Reliability and Validity

We assessed the reliability and validity of the latent constructs with confirmatory factor analysis (CFA). The model fit indices were Chi-square (125) = 605.22, RMSEA = 0.064, standardized RMR = 0.036, CFI = 0.95 and NNFI = 0.93, indicating that the model was acceptable and confirming convergent validity. As

Table 5.1 Descriptive statistics and correlations

	1	2	3	4	5	6	7	8	9	10	11	12	13
1. HR													
2. DMT	0.36*												
3. Internal integration	0.39*	0.34*											
4. Supplier integration	0.44*	0.45*	0.55*										
5. Customer integration	0.36*	0.50*	0.46*	0.67*									
6. Competition	0.08*	0.09*	0.10*	0.14*	0.16*								
7. Industry 1	−0.12*	−0.09*	−0.05	−0.17*	−0.0187	0.01							
8. Industry 2	0.07*	0.08*	0.06	0.10*	0.12*	0.04*	−0.25*						
9. Industry 3	0.06	0.07*	0.06	0.10*	0.04	0.05	−0.29*	−0.17*					
10. Industry 4	−0.04	−0.11*	−0.12*	−0.06	−0.15*	−0.09*	−0.38*	−0.22*	−0.25*				
11. Industry 5	0.07*	0.12*	0.06	0.12*	0.08*	0.02	−0.22*	−0.13*	−0.15*	−0.19*			
12. Industry 6	0.05	−0.01	0.04	−0.01	−0.03	−0.02	−0.16*	−0.09*	−0.10*	−0.14*	−0.08*		
13. Firm size	0.13*	0.18*	0.11*	0.20*	0.16*	0.04*	−0.14*	0.08*	0.00	−0.05	0.17*	0.05	
Mean	3.419	2.834	3.554	3.145	3.059	3.304	0.303	0.132	0.164	0.248	0.100	0.053	6.028
SD	0.765	1.031	0.859	0.840	0.965	0.664	0.460	0.339	0.371	0.432	0.300	0.223	1.720

Note $*p < 0.05$

5.3 Method 79

Table 5.2 CFA results

Construct	Item	Factor loading	CR	AVE
HR	• HR1	0.77	0.73	0.48
	• HR2	0.71		
	• HR3	0.58		
DMT	• DMT1	0.66	0.80	0.57
	• DMT2	0.84		
	• DMT3	0.75		
Internal integration	• II1	0.76	0.89	0.67
	• II2	0.79		
	• II3	0.86		
	• II4	0.86		
Supplier integration	• SI1	0.76	0.84	0.58
	• SI2	0.82		
	• SI3	0.78		
	• SI4	0.67		
Customer integration	• CI1	0.84	0.88	0.65
	• CI2	0.85		
	• CI3	0.78		
	• CI4	0.76		

Table 5.2 shows, the composite reliability (CR) of each latent construct exceeded 0.70. To assess discriminant validity, we compared the square root of AVE with the construct's correlations. All the square roots of AVEs were greater than the correlations, providing support of discriminant validity.

5.4 Results

We used ordinary least squares (OLS) regression to test the proposed hypotheses. The results were presented in Table 5.3. The results of Models 1, 3, and 5 indicated that the relationships between HR and the three dimensions of SCI were positive and significant, supporting H1a–c. The results of Models 1, 3, and 5 further revealed that the relationships between DMT and the three dimensions of SCI were all positive and significant, supporting H2a–c.

As for the moderating effects of competition, the results of Model 2 showed that the interaction of HR and the competition was negatively related to internal integration. Thus, H3a was rejected. The results of Model 4 showed that the interaction of HR and competition was not significantly related to supplier integration, offering no support for H3b. The results of Model 6 indicated that the interaction effect of HR

Table 5.3 Regression results

	Dependent variables					
	Internal integration		Supplier integration		Customer integration	
	M1	M2	M3	M4	M5	M6
Constant	1.84^{***}	1.57^{***}	0.92^{***}	0.54^{**}	0.99^{***}	0.51^{*}
	(0.14)	(0.14)	(0.13)	(0.18)	(0.15)	(0.21)
Controls						
Industry2	0.03	0.02	0.23^{**}	0.23^{**}	0.10	0.11
	(0.08)	(0.08)	(0.08)	(0.08)	(0.09)	(0.09)
Industry3	0.03	0.03	0.23^{**}	0.22^{**}	-0.07	-0.08
	(0.08)	(0.08)	(0.07)	(0.07)	(0.08)	(0.08)
Industry4	-0.14^{*}	-0.13^{*}	0.11^{a}	0.12^{*}	-0.24^{**}	-0.21^{**}
	(0.07)	(0.07)	(0.06)	(0.06)	(0.07)	(0.07)
Industry5	0.01	0.02	0.23^{**}	0.22^{**}	-0.05	-0.06
	(0.09)	(0.09)	(0.09)	(0.09)	(0.10)	(0.10)
Industry6	0.10	0.09	0.03	0.05	-0.22^{a}	-0.18
	(0.12)	(0.12)	(0.11)	(0.11)	(0.13)	(0.13)
Size	0.01	0.01	0.04^{**}	0.04^{**}	0.03^{*}	0.03^{*}
	(0.02)	(0.02)	(0.01)	(0.01)	(0.02)	(0.02)
Independent variables						
HR	0.34^{***}	0.33^{***}	0.33^{***}	0.33^{***}	0.26^{***}	0.25^{***}
	(0.04)	(0.04)	(0.03)	(0.03)	(0.04)	(0.04)
DMT	0.18^{***}	0.18^{***}	0.25^{***}	0.25^{***}	0.38^{***}	0.37^{***}
	(0.03)	(0.03)	(0.02)	(0.02)	(0.03)	(0.03)
Competition		0.09^{*}		0.12^{**}		0.15^{**}
		(0.05)		(0.04)		(0.05)
Interactions						
HR $\times$ Competition		-0.14^{*}		0.06		0.13^{*}
		(0.06)		(0.05)		(0.06)
DMT $\times$ Competition		0.03		0.02		0.00
		(0.04)		(0.04)		(0.04)
R^2	0.203	0.211	0.315	0.323	0.301	0.314

Note The standard errors for each unstandardized parameter estimate are shown in parentheses
$^{a}p < 0.1$, $^{*}p < 0.05$, $^{**}p < 0.01$, $^{***}p < 0.001$

and competition on customer integration was significantly positive, supporting H3c. The results of Models 2, 4 and 6 also indicated that the interaction effects of DMT and competition on the internal, supplier, and customer integration are non-significant, thus rejecting H4a, H4b, and H4c.

5.4 Results

To better understand the significant moderating effects, we plotted the interaction effect at one standard deviation below the mean, and one standard deviation above the mean of the moderator. Specifically, Fig. 5.2(a) demonstrated that the relationship between HR and internal integration changed from positive to negative as competition increased. Figure 5.2(b) demonstrated the relationship between HR and customer integration was more positive when the competition is high.

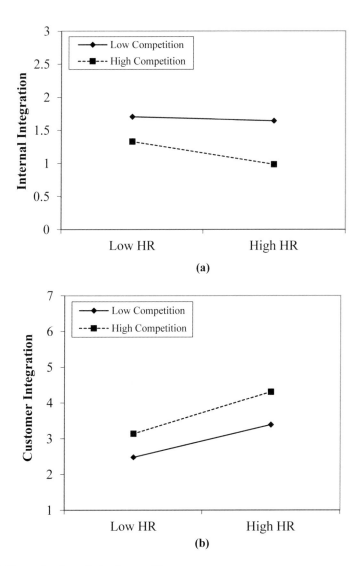

Fig. 5.2 The moderating effects of competition

5.5 Discussion

5.5.1 Findings

From the STS perspective, this study identified two important internal enablers (i.e., DMT and HR) that a manufacturing firm may use to promote the three dimensions of SCI and an environmental factor (i.e., competition) that may influence the effectiveness of HR. Our results revealed that (1) both DMT and HR have positive effects on all three dimensions of SCI; and (2) competition negatively moderates the effect of HR on internal integration but positively moderates the effect of HR on customer integration.

Our results suggest that HR are significantly related to the three dimensions of SCI, which demonstrates the significant role that employees' involvement plays in SCI, consistent with previous studies of Huo et al. (2015a) and Xu et al. (2014). The implementation of SCI largely depends on the engagement of staff, who must complete specific tasks and identify the firm's common norms and values. Employees with multiple skills can facilitating the integration of internal functions (Zhao et al., 2011). Furthermore, employees with a clear understanding of SCI and a strong desire to achieve corporate goals promote information sharing and communication among internal functions and maintain good partnerships with external partners (Huo et al., 2016b). Our results also reveal that DMT is significantly related to the three dimensions of SCI. Advanced digital technologies provide easy access to the huge amount of operational data for both internal members and supply chain partners.

Although both HR and DMT are vital in the implementation of SCI, their effects are differently affected by competition. Specifically, competition can influence the role of HR in enhancing customer integration but does not influence the role of DMT in any integration. We offer two reasons for this difference. First, it may attribute to the difference between those two factors. Specifically, compared with technical resources, HR has a sense and perception of the environment. As the competition increases, employees will generate a sense of crisis, with which they will make use of their subjective initiative and play their roles to promote the implementation of corporate strategy. Second, DMT mainly focuses on the manufacturing processes. SCI will take advantages of DMT's side effects as information acquirement and sharing, which are relative stable procedures that cannot be changed or updated easily. If the transparency of supply chains and flexibility of job shops are once enhanced by DMT, they will be exploited as much as they could and cannot be further improved unless DMT is further advanced, which may have little connection with different levels of present competition.

Besides, we find that the moderating effects of competition on the HR–SCI relationship are different in terms of different SCI dimensions. The difference mainly arises from the essential difference among different SCI dimensions and the nature of HR. Specifically, customer integration requires coordination work with other firms, the complexity and difficulty of which are larger than coordinating internal functions. When competition increases, inter-firm coordination will become more challenging. In this way, promoting customer integration is an especially challenging task for employees. However, people's potential and creativity can be stimulated by challenging and complex tasks (Chang et al., 2014). Besides, as the manufacturer faces resource limitations, it cannot simultaneously promote all the three dimensions of SCI. In this way, the manufacturer will allocate its resource in the most critical activity. Since customers are especially important in ensuring the firm's financial performance and determining the success in converting inputs into outputs, the motivation of customer integration is highlighted. High-level competition requires fast response to customers' changing needs. Therefore, it is a rational choice for employees to devote more efforts to improve customer integration. However, considering that supplier integration cannot bring in profits directly (Flynn et al., 2010), a firm's motivation for supplier integration can hardly be stimulated. Besides, the knowledge and skills of organizations may pose more pressure to suppliers under extensive competition, which may counteract the effort of their employees.

In addition, contrary to our hypothesis, competition negatively moderates the effect of HR on internal integration, which indicates that when competition is high, the effectiveness of HR will be mitigated. We offer several possible reasons for this finding. First, as high competition will generate uncertainties and risks, employees may pay more attention to dealing with the external environments, resulting a lack of focus on the issues within the firm. Second, as a firm is facing threats of external competition, there will appear conflicts of interests among different departments within the firm. In this way, employees from different departments who realize the conflicts are less likely to ally and coordinate the actions. Therefore, when competition is high, employees' motivation and identification for internal integration will be lessened.

5.5.2 Theoretical Implications

Based on the STS perspective, this study examines the roles of social subsystem (i.e., HR) and technical subsystem (i.e., DMT) on the three dimensions of SCI and the moderating role of an environmental factor (i.e., competition). This study contributes to the literature in several ways.

This study contributes to the literature on the enablers of SCI by identifying the three significant factors based on the STS perspective. First, in this study, HR represents the social factor, which is regarded as a soft resource, while DMT is considered to be technical capabilities, which is regarded as hard resources. Although the two resources improve all three dimensions of SCI, their effects are differently affected by competition, which is regarded as an environmental factor. By involving competition, this study further illustrates the different effects of those social and technical factors in promoting SCI. In the meantime, supplier integration is not moderated and internal integration is negatively influenced. Second, this study enriches the understanding of the difference between internal, supplier, and customer integration in terms of the importance and difficulty of implementation by revealing that competition differently moderates the effects of HR on different dimensions of SCI.

This study also contributes to the DMT literature. The emerging technologies in Industry 4.0 have been increasingly used by manufacturers in their manufacturing processes. However, there is a lack of empirical studies that investigate the antecedents and consequence of DMT. This study provides empirical supports by validating the effective role of DMT in facilitating the three dimensions of SCI. It indicates that the manufacturer's pursuit of the most advanced manufacturing technologies can also benefit the firm's strategic activities even if the activities are cross-boundary to organizations. This study provides an unexpected result that DMT cannot be moderated by competition, which deserves further investigation.

In addition, the study also contributes to the STS literature by introducing the overlooked environmental side of the STS in examining the enablers of SCI. The STS perspective is widely used to design and explain how organizations achieve outstanding capabilities and superior performance with the optimal interactions across subsystems. SCI, as a firm's strategic practice, is especially significant in today's environment characterized by quickly updating technologies and intensive competition. Considering that the manufacturer's motivation and the effectiveness of implementing SCI are influenced by factors from different aspects, including different resources and environmental factors, the all-sided investigation from the STS perspective provides a more integrated guide.

5.6 Conclusions

Based on the STS perspective, we propose and test a model of relationships between DMT, HR, SCI, and competition in the context of global manufacturing. Our results indicate that DMT and HR play significant roles in implementing the three dimensions of SCI, and the effects of HR are affected differently by competition. This study not only fills the gap of the positive effects of DMT on SCI, but also provides complementary empirical supports by better portraying the roles of social and technical factors under the influence of an environmental factor (i.e., competition).

Although our study provides several significant contributions, there also exist limitations. First, we consider DMT and HR as enablers of SCI, but we do not measure the match or alignment between them. Future studies can identify the role of the match between social and technical factors in promoting SCI. Second, whether DMT can have better influence on SCI and other supply chain practices under certain circumstance is worth studying. Finally, we use cross-sectional data to test the conceptual model, which does not examine causal relationships. Future research could use panel data to explore causal relationships between the constructs examined in this study.

References

Agrifoglio, R., Cannavale, C., Laurenza, E., & Metallo, C. (2017). How emerging digital technologies affect operations management through co-creation. Empirical evidence from the maritime industry. *Production Planning & Control, 28*(16), 1298–1306.

Alfalla-Luque, R., Marin-Garcia, J. A., & Medina-Lopez, C. (2015). An analysis of the direct and mediated effects of employee commitment and supply chain integration on organisational performance. *International Journal of Production Economics, 162*, 242–257.

Alfalla-Luque, R., Medina-Lopez, C., & Dey, P. K. (2013). Supply chain integration framework using literature review. *Production Planning & Control, 24*(8–9), 800–817.

Aversa, P., Formentini, M., Iubatti, D., & Lorenzoni, G. (2021). Digital machines, space, and time: Towards a behavioral perspective of flexible manufacturing. *Journal of Product Innovation Management, 38*(1), 114–141.

Bae, H. S. (2017). Empirical relationships of perceived environmental uncertainty, supply chain collaboration and operational performance: Analyses of direct, indirect and total effects. *Asian Journal of Shipping and Logistics, 33*(4), 263–272.

Batt, R. (2002). Managing customer services: Human resource practices, quit rates, and sales growth. *Academy of Management Journal, 45*(3), 587–597.

Birasnav, M., & Bienstock, J. (2019). Supply chain integration, advanced manufacturing technology, and strategic leadership: An empirical study. *Computers & Industrial Engineering, 130*, 142–157.

Bokrantz, J., Skoogh, A., Berlin, C., & Stahre, J. (2017). Maintenance in digitalised manufacturing: Delphi-based scenarios for 2030. *International Journal of Production Economics, 191*, 154–169.

Bostrom, R. P., & Heinen, J. S. (1977). MIS problems and failures: A socio-technical perspective Part I: The causes. *MIS Quarterly, 1*(3), 17–32.

Boyer, K. K., Leong, G. K., Ward, P. T., & Krajewski, L. J. (1997). Unlocking the potential of advanced manufacturing technologies. *Journal of Operations Management, 15*(4), 331–347.

Buer, S. V., Strandhagen, J. W., Semini, M., & Strandhagen, J. O. (2021). The digitalization of manufacturing: Investigating the impact of production environment and company size. *Journal of Manufacturing Technology Management, 32*(3), 621–645.

Chang, S., Jia, L., Takeuchi, R., & Cai, Y. (2014). Do high-commitment work systems affect creativity? A multilevel combinational approach to employee creativity. *Journal of Applied Psychology, 99*(4), 665.

Chen, H., Daugherty, P. J., & Landry, T. D. (2009). Supply chain process integration: A theoretical framework. *Journal of Business Logistics, 30*(2), 27–46.

Chaudhuri, A., & Jayaram, J. (2019). A socio-technical view of performance impact of integrated quality and sustainability strategies. *International Journal of Production Research, 57*(5), 1478–1496.

Closs, D. J., Jacobs, M. A., Swink, M., & Webb, G. S. (2008). Toward a theory of competencies for the management of product complexity: Six case studies. *Journal of Operations Management, 26*(5), 590–610.

Cohen, W. M., & Levinthal, D. A. (1990). Absorptive capacity: A new perspective on learning and innovation. *Administrative Science Quarterly, 35*(1), 128–152.

Dalenogare, L. S., Benitez, G. B., Ayala, N. F., & Frank, A. G. (2018). The expected contribution of Industry 4.0 technologies for industrial performance. *International Journal of Production Economics, 204*, 383–394.

DiMaggio, P. J., & Powell, W. W. (1983). The iron cage revisited: Institutional isomorphism and collective rationality in organizational fields. *American Sociological Review, 48*(2), 147–160.

Van Donk, D. P., Sancha, C., & Scholten, K. (2017). Does supply chain integration help or hinder in building resilient supply chains? *Academy of Management Proceedings, 1*, 13010.

Emery, F. (1959). *Characteristics of socio-technical systems.* Tavistock Institute.

Ellinger, A. E., & Ellinger, A. D. (2014). Leveraging human resource development expertise to improve supply chain managers' skills and competencies. *European Journal of Training and Development, 38*(1–2), 118–135.

Ellinger, A. E., Keller, S. B., & Baş, A. B. E. (2010). The empowerment of frontline service staff in 3PL companies. *Journal of Business Logistics, 31*(1), 79–98.

Flynn, B. B., Huo, B., & Zhao, X. (2010). The impact of supply chain integration on performance: A contingency and configuration approach. *Journal of Operations Management, 28*(1), 58–71.

Frank, A. G., Dalenogare, L. S., & Ayala, N. F. (2019). Industry 4.0 technologies: Implementation patterns in manufacturing companies. *International Journal of Production Economics, 210*, 15–26.

Fujitsu Components America. (2021). *Digital integration of manufacturing supply and engineering chains is vital for competitive success.* https://www.automation.com/en-us/articles/april-2021/digital-integration-manufacturing-supply-chains

Ganbold, O., Matsui, Y., & Rotaru, K. (2020). Effect of information technology-enabled supply chain integration on firm's operational performance. *Journal of Enterprise Information Management, 34*(3), 948–989.

Geissbauer, R., Schrauf, S., Berttram, P., & Cheraghi, F. (2017). *Digital factories 2020: Shaping the future manufacturing.* Berlin, Germany. https://www.pwc.de/de/digitale-transformation/digital-factories-2020-shaping-the-future-of-manufacturing.pdf

Gillani, F., Chatha, K. A., Sadiq Jajja, M. S., & Farooq, S. (2020). Implementation of digital manufacturing technologies: Antecedents and consequences. *International Journal of Production Economics, 229*, 107748.

Handley, S. M., & Benton, W. C. (2012). The influence of exchange hazards and power on opportunism in outsourcing relationships. *Journal of Operations Management, 30*(1), 55–68.

Himmler, F. (2014). The digital factory: A reference process based software market analysis. *International Journal of Distributed Systems and Technologies, 5*(2), 17–30.

Hitt, M. A., Hoskisson, R. E., & Nixon, R. D. (1993). A mid-range theory of interfunctional integration, its antecedents and outcomes. *Journal of Engineering and Technology Management, 10*(1–2), 161–185.

Hofmann, E., & Rüsch, M. (2017). Industry 4.0 and the current status as well as future prospects on logistics. *Computers in Industry, 89*, 23–34.

Holmström, J., Holweg, M., Lawson, B., Pil, F. K., & Wagner, S. M. (2019). The digitalization of operations and supply chain management: Theoretical and methodological implications. *Journal of Operations Management, 65*(8), 728–734.

Holmström, J., & Partanen, J. (2014). Digital manufacturing-driven transformations of service supply chains for complex products. *Supply Chain Management, 19*(4), 421–430.

Huo, B., Han, Z., Chen, H., & Zhao, X. (2015a). The effect of high-involvement human resource management practices on supply chain integration. *International Journal of Physical Distribution & Logistics Management, 45*(8), 716–746.

Huo, B., Han, Z., & Prajogo, D. (2016a). Antecedents and consequences of supply chain information integration: A resource-based view. *Supply Chain Management, 21*(6), 661–677.

Huo, B., Ye, Y., Zhao, X., & Shou, Y. (2016b). The impact of human capital on supply chain integration and competitive performance. *International Journal of Production Economics, 178*, 132–143.

Huo, B., Zhang, C., & Zhao, X. (2015b). The effect of IT and relationship commitment on supply chain coordination: A contingency and configuration approach. *Information & Management, 52*(6), 728–740.

Jacobs, M. A., Yu, W., & Chavez, R. (2016). The effect of internal communication and employee satisfaction on supply chain integration. *International Journal of Production Economics, 171*, 60–70.

Jajja, M. S. S., Chatha, K. A., & Farooq, S. (2018). Impact of supply chain risk on agility performance: Mediating role of supply chain integration. *International Journal of Production Economics, 205*, 118–138.

References

Kamble, S. S., Gunasekaran, A., & Sharma, R. (2018). Analysis of the driving and dependence power of barriers to adopt Industry 4.0 in Indian manufacturing industry. *Computers in Industry, 101*, 107–119.

Kathuria, R. (2000). Competitive priorities and managerial performance: A taxonomy of small manufacturers. *Journal of Operations Management, 18*(6), 627–641.

Kull, T. J., Ellis, S. C., & Narasimhan, R. (2013). Reducing behavioral constraints to supplier integration: A socio-technical systems perspective. *Journal of Supply Chain Management, 49*(1), 64–86.

Lengnick-Hall, M. L., Lengnick-Hall, C. A., & Rigsbee, C. M. (2013). Strategic human resource management and supply chain orientation. *Human Resource Management Review, 23*(4), 366–377.

Liao, Y., Deschamps, F., Loures, E. d. F. R., & Ramos, L. F. P. (2017). Past, present and future of Industry 4.0—A systematic literature review and research agenda proposal. *International Journal of Production Research, 55*(12), 3609–3629.

Liker, J. K., Kamath, R. R., & Nazli Wasti, S. (1998). Supplier involvement in design: A comparative survey of automotive suppliers in the USA, UK and Japan. *International Journal of Quality Science, 3*(3), 214–238.

Liu, H., Huang, Q., Wei, S., & Huang, L. (2015). The impacts of IT capability on internet-enabled supply and demand process integration, and firm performance in manufacturing and services. *International Journal of Logistics Management, 26*(1), 172–194.

Lorenz, R., Benninghaus, C., Friedli, T., & Netland, T. H. (2020). Digitization of manufacturing: The role of external search. *International Journal of Operations & Production Management, 40*(7–8), 1129–1152.

MacDuffie, J. P. (1995). Human resource bundles and manufacturing performance: Organizational logic and flexible production systems in the world auto industry. *Industrial & Labor Relations Review, 48*(2), 197–221.

Manz, C. C., & Stewart, G. L. (1997). Attaining flexible stability by integrating total quality management and socio-technical systems theory. *Organization Science, 8*(1), 59–70.

McAfee, R. B., Glassman, M., & Earl, D. H., Jr. (2002). The effects of culture and human resource management policies on supply chain management strategy. *Journal of Business Logistics, 23*(2), 1–18.

Menon, S. T. (2012). Human resource practices, supply chain performance, and wellbeing. *International Journal of Manpower, 33*(7), 769–785.

Moyano-Fuentes, J., Sacristán-Díaz, M., & Garrido-Vega, P. (2016). Improving supply chain responsiveness through advanced manufacturing technology: The mediating role of internal and external integration. *Production Planning & Control, 27*(9), 686–697.

Narayanan, S., Jayaraman, V., Luo, Y., & Swaminathan, J. M. (2011). The antecedents of process integration in business process outsourcing and its effect on firm performance. *Journal of Operations Management, 29*(1–2), 3–16.

Oliveira, T., Martins, R., Sarker, S., Thomas, M., & Popovič, A. (2019). Understanding SaaS adoption: The moderating impact of the environment context. *International Journal of Information Management, 49*, 1–12.

Pagell, M. (2004). Understanding the factors that enable and inhibit the integration of operations, purchasing and logistics. *Journal of Operations Management, 22*(5), 459–487.

Pasmore, W., Francis, C., Haldeman, J., & Shani, A. (1982). Sociotechnical systems: A North American reflection on empirical studies of the seventies. *Human Relations, 35*(12), 1179–1204.

Prajogo, D., & Olhager, J. (2012). Supply chain integration and performance: The effects of long-term relationships, information technology and sharing, and logistics integration. *International Journal of Production Economics, 135*(1), 514–522.

Shou, Y., Kang, M., & Park, Y. W. (2022). *Supply chain integration for sustainable advantages.* Springer.

Shou, Y., Li, Y., Park, Y. W., & Kang, M. (2017). The impact of product complexity and variety on supply chain integration. *International Journal of Physical Distribution & Logistics Management, 47*(4), 297–317.

Shou, Y., Zhao, X., Dai, J., & Xu, D. (2021). Matching traceability and supply chain coordination: Achieving operational innovation for superior performance. *Transportation Research Part E: Logistics and Transportation Review, 145*, 102181.

Siawsh, N., Peszynski, K., Young, L., & Vo-Tran, H. (2021). Exploring the role of power on procurement and supply chain management systems in a humanitarian organisation: A socio-technical systems view. *International Journal of Production Research, 59*(12), 3591–3616.

Sodhi, M. S., & Tang, C. S. (2019). Research opportunities in supply chain transparency. *Production and Operations Management, 28*(12), 2946–2959.

Strozzi, F., Colicchia, C., Creazza, A., & Noè, C. (2017). Literature review on the 'Smart Factory' concept using bibliometric tools. *International Journal of Production Research, 55*(22), 6572–6591.

Szalavetz, A. (2019). Industry 4.0 and capability development in manufacturing subsidiaries. *Technological Forecasting & Social Change, 145*, 384–395.

Teller, C., Kotzab, H., & Grant, D. B. (2012). Improving the execution of supply chain management in organizations. *International Journal of Production Economics, 140*(2), 713–720.

Venkatesh, V., & Bala, H. (2012). Adoption and impacts of interorganizational business process standards: Role of partnering synergy. *Information Systems Research, 23*(4), 1131–1157.

Vinodh, S., Sundararaj, G., Devadasan, S. R., Kuttalingam, D., & Rajanayagam, D. (2009). Agility through rapid prototyping technology in a manufacturing environment using a 3D printer. *Journal of Manufacturing Technology Management, 20*(7), 1023–1041.

Waldman, D. A. (1994). The contributions of total quality management to a theory of work performance. *Academy of Management Review, 19*(3), 510–536.

Wang, K., Huo, B., & Tian, M. (2021) How to protect specific investments from opportunism: A moderated mediation model of customer integration and transformational leadership. *International Journal of Production Economics, 232*, 107938.

Wang, Z., Huo, B., Qi, Y., & Zhao, X. (2016). A resource-based view on enablers of supplier integration: Evidence from China. *Industrial Management & Data Systems, 116*(3), 416–444.

Wright, P. M., McMahan, G. C., & McWilliams, A. (1994). Human resources and sustained competitive advantage: A resource-based perspective. *International Journal of Human Resource Management, 5*(2), 301–326.

Xu, D., Huo, B., & Sun, L. (2014). Relationships between intra-organizational resources, supply chain integration and business performance. *Industrial Management & Data Systems, 114*(8), 1186–1206.

Xu, L. D., Xu, E. L., & Li, L. (2018). Industry 4.0: state of the art and future trends. *International Journal of Production Research, 56*(8), 2941–2962.

Yin, Y., Stecke, K. E., & Li, D. (2018). The evolution of production systems from Industry 2.0 through Industry 4.0. *International Journal of Production Research, 56*(1–2), 848–861.

Zhang, Y., Zhao, X., & Huo, B. (2019). The impacts of intra-organizational structural elements on supply chain integration. *Industrial Management & Data Systems, 119*(5), 1031–1045.

Zhao, L., Huo, B., Sun, L., & Zhao, X. (2013). The impact of supply chain risk on supply chain integration and company performance: A global investigation. *Supply Chain Management, 18*(2), 115–131.

Zhao, X., Huo, B., Selen, W., & Yeung, J. H. Y. (2011). The impact of internal integration and relationship commitment on external integration. *Journal of Operations Management, 29*(1), 17–32.

Zhou, H., Keong Leong, G., Jonsson, P., & Sum, C. C. (2009). A comparative study of advanced manufacturing technology and manufacturing infrastructure investments in Singapore and Sweden. *International Journal of Production Economics, 120*(1), 42–53.

References

Zsidisin, G. A., Hartley, J. L., Bernardes, E. S., & Saunders, L. W. (2015). Examining supply market scanning and internal communication climate as facilitators of supply chain integration. *Supply Chain Management, 20*(5), 549–560.

Chapter 6
Risk Management of Manufacturing Multinational Corporations: The Effects of Supply Chain Integration

Abstract Through the lens of organizational information processing theory, this study aims to investigate the contingencies of supply chain risk management (SCRM) in manufacturing multinational corporations (MNCs) by exploring the moderating role of international asset dispersion (IAD) in the performance effect of SCRM as well as the counteraction effect of supply chain integration (SCI). Survey data from a sample of manufacturing MNCs were analyzed. Hierarchical regression analysis was conducted to test the proposed hypotheses. The results demonstrate that SCRM improves the operational effectiveness of manufacturing MNCs but this performance effect is attenuated by IAD. Nevertheless, external integration can counteract the negative effect of IAD and ensure the efficacy of SCRM practices. This study sheds light on the SCRM-operational performance relationship by considering how a manufacturing MNC's IAD may influence the efficacy of SCRM practices and how SCI can attenuate the negative effect of IAD.

Keywords Supply chain integration · Supply chain risk management · Manufacturing multinational corporation · International asset dispersion

6.1 Introduction

The relentless advance of globalization and international trade has created an interconnected supply chain that, while rich in opportunities, is vulnerable to disruption. Disruptions can be global geopolitical events, such as Brexit and the steel tariffs proposed by President Trump of the United States; or they can be local natural disasters, such as the Kumamoto earthquake in Japan that forced the automotive industry to suspend production and wildfires in California that cut off key rail and trucking routes. As a result, supply chain risk management (SCRM), which is defined as the practices that can prevent, detect and mitigate operations and supply chain risks, is widely adopted by manufacturing firms to cope with the increased complexity and

---

This chapter is a revised version of the following journal paper: Hu, W., Shou, Y., Kang, M., & Park, Y. (2020). Risk management of manufacturing multinational corporations: the moderating effects of international asset dispersion and supply chain integration. *Supply Chain Management: An International Journal*, 25(1), 61–76.

© The Author(s), under exclusive license to Springer Nature Singapore Pte Ltd. 2022
Y. Shou et al., *Supply Chain Integration for Sustainable Advantages*,
https://doi.org/10.1007/978-981-16-9332-8_6

risks (Chaudhuri et al., 2018; Kauppi et al., 2016; Manuj et al., 2014; Munir et al., 2020). However, the existing findings on the impact of SCRM practices on firm performance are mixed (Colicchia & Strozzi, 2012; Kauppi et al., 2016; Manuj et al., 2014; Wiengarten et al., 2016). Thus, whether and how SCRM practices can improve operational performance remains unclarified and requires further investigation.

With the presence of the global manufacturing networks (GMNs), while multinational corporations (MNCs) have enhanced their competitiveness, they also face higher level of uncertainty. According to a study by the Chartered Institute of Procurement and Supply (CIPS), customs delays due to Brexit could bankrupt 10% of U.K. businesses that have European Union (EU) suppliers and almost two-thirds of EU businesses that work with suppliers located in the U.K. expect to move some part of their supply chains out of the U.K. (Green, 2017). Post Brexit, many multinationals that depend heavily on imports from the EU and are closely integrated with EU-based suppliers face considerable uncertainty. This uncertainty will not be short-lived: it could last a decade or more, and will add to existing volatility and vulnerability of supply chains in various industries. According to McKinsey' interviews with U.K.-based executives, deep concerns over the impact of Brexit are found: advanced manufacturing companies are worried about the impact of tariffs, warehousing and border friction on existing supply chains, especially in just-in-time (JIT) manufacturing (Gysegom et al., 2019). In contrast with the prevalence of manufacturing MNCs, the influences of contextual factors in international business field have not been fully examined in the extant operations and supply chain management literature. In this study, we extend the risk management research by considering the impact of a manufacturing MNC's international asset dispersion (IAD) on the efficacy of SCRM practices. Broader IAD tends to increase the spatial and environmental complexity of a manufacturing MNC (Bode & Wagner, 2015; Tang, 2006), which in turn delays the collection and obfuscates the accuracy of operations and supply chain information (Goerzen & Beamish, 2003; Zaheer & Hernandez, 2011). Since SCRM is an information-intensive process whose function is to deal with all kinds of uncertainty (Srinivasan & Swink, 2015; Swink et al., 2007) and obtain accurate information on internal operations and external supply chains to identify, assess, mitigate and monitor risks (Kauppi et al., 2016; Manuj & Mentzer, 2008b; Sodhi et al., 2012). IAD, as one of the major sources of environmental uncertainty, may influence the efficacy of SCRM practices. However, to our best knowledge, there is no empirical study that has investigated the moderating effect of IAD on the relationship between SCRM practices and operational performance.

Some studies have suggested that firms can develop capabilities and strategies to overcome the possible threat of different types of supply chain risks to the efficacy of SCRM practices (Manuj et al., 2014). In fact, some MNCs have adopted a variety of strategies to deal with problems caused by supply chain uncertainties. For example, Intel, as one of the world's largest and highest valued semiconductor chip makers, has five wafer fabs in production worldwide at ten locations (Intel, 2020a) and a decentralized supply chain with 9000 tier 1 suppliers in 89 countries (Intel, 2020b). The large and diverse group of suppliers added great uncertainty to Intel's supply chain network. After setting up a supply chain management (SCM) department, Intel implemented

6.1 Introduction

the enterprise resource planning (ERP) system throughout the company to integrate the internal supply chain within the enterprise. Although the internal supply chain integration (SCI) was completed, there was no timely information sharing between the enterprise, suppliers and customers, causing problems such as inventory over-stocking, shutdown, delayed delivery and etc. Therefore, Intel extended the internal SCI to external relationships and established strategic partnerships with upstream and downstream firms (Brown, 2015). The practices that Intel implemented to manage upstream suppliers include developing long-term relationships with top-tier suppliers who have proven track records for delivering raw materials of high quality. Besides, Intel also placed emphasis on supporting suppliers through specialized training on business skills, human resources, environmental management and so on. In order to achieve this, Intel Partner University was launched to help suppliers to grow their expertise on various topics, solutions and specialties. For downstream customers, Intel deployed a customer relationship management (CRM) system, from which Intel can obtain customer inventory status and material consumption forecast. Through the internal and external SCI, Intel not only optimized the internal management process but also realized timely and reliable external information sharing.

Through the case of Intel, it is observed that SCI offers the potential to manage the possible threat of IAD by providing rich and accurate operations and supply chain information (Kleindorfer & Saad, 2009; Leuschner et al., 2013; Sodhi et al., 2012). Although there are wide discussion on the significant role of SCI in improving firm performance in high risk situations (Braunscheidel & Suresh, 2009; Ganbold et al., 2021; Prajogo & Olhager, 2012; Schoenherr & Swink, 2012; Wieland & Wallenburg, 2013), no empirical study has investigated the role of SCI in mitigating the potential threat of IAD. In this study, we intend to narrow this gap by examining the essential role of SCI in counteracting the negative effect of IAD on the relationship between SCRM practices and operational performance. Therefore, the research questions of this study are as follows:

RQ1 What is the effect of SCRM practices on operational performance?

RQ2 How does IAD moderate the efficacy of SCRM practices on operational performance?

RQ3 How and to what extent does SCI counteract the effect of IAD on the relationship between SCRM practices and operational performance?

In this chapter, organizational information processing theory (OIPT) is applied to develop hypotheses. This study contributes to the SCM literature by distinguishing the effects of SCRM practices on individual dimensions of operational performance, providing in-depth explanation for the mixed findings on the efficacy of SCRM practices in the extant literature. Besides, this study extends the research scope of SCM by linking a construct in international business research (i.e., IAD) to SCRM and SCI of manufacturing MNCs. Specifically, this study investigates the attenuating effect of IAD and then the counteracting effect of SCI on the efficacy of SCRM practices of manufacturing MNCs.

The rest of this chapter is organized as follows. In Sect. 6.2, we review the OIPT literature and then develop our hypotheses. Section 6.3 introduces the data and variables used in this study. Section 6.4 presents the data analysis results. Section 6.5 provides the major findings and discusses the theoretical contributions of this research. Finally, we draw conclusions in Sect. 6.6.

6.2 Literature Review and Hypotheses Development

6.2.1 Organizational Information Processing Theory

Galbraith (1974a) first proposed OIPT and argued that each organization is an information processing system that can deal with uncertainty either by reducing information processing needs or by enhancing information processing capabilities. From the perspective of OIPT, the information processing requirements are driven by uncertainty, which refers to "the difference between the amount of information required to perform the task and the amount of information already possessed by the organization" (Galbraith, 1974b), while the information processing capacity is reflected by organizational design strategies such as organizational structure or organization process. Tushman and Nadler (1978) further suggested that companies could be more effective if their information processing capability fits their information processing requirements. Hence, as the degree of uncertainty increases, organizations have to enhance their capabilities in information collection, information processing and information transmission.

A firm's information processing needs refer to the communication requirements for inter-organizational interactions (Premkumar et al., 2005). Information processing needs are captured by uncertainty, the sources of which are divided into three types: task characteristics, task environment and inter-organizational interdependence (Tushman & Nadler, 1978). Scholars have explored the task characteristics from different aspects such as task predictability, task complexity and task analyzability (Daft & Lengel, 1986; Perrow, 1967; Tushman & Nadler, 1978; Ven et al., 1976), all of which have a great influence on the level of uncertainty. Although only few studies focusing on inter-organizational interdependence have been conducted, there is evidence that the more intricate the task interdependence, the greater the uncertainty associated with the task. Stemming from the complexity and instability of the environment, environmental uncertainty is the most widely discussed factor that causes an organization's information processing needs. In general, the more dynamic the environment, the greater the uncertainty faced by the organization. Typically, an organization can reduce its information processing needs by creating slack resources, redundancies or self-contained tasks, such as building inventory buffers, increasing the budgets etc. However, such strategies are costly and may damage the firm's competitiveness.

On the other hand, a firm's information processing capacity refers to the ability to act on the information collected, including the gathering, organization, and exploitation of new information, as well as the ability to utilize the information to support

6.2 Literature Review and Hypotheses Development

business operations (Galbraith, 1974a). The organization can increase its information processing capacity by investing in both lateral relations and vertical information systems (Srinivasan & Swink, 2015). The lateral relations include the organizational processes and relations that enable organizations to gather valuable information from suppliers, customers and different departments within the organization. These lateral relationships, while improving the richness of information by creating additional information exchange, also decrease the information processing needs by reducing the equivocality in the information. Vertical information systems, another strategy proposed by Galbraith (1974a) to enhance information processing capacity, enable organizations to process information during task performance in ways that do not overload hierarchical communication channels. Vertical information systems enable organizations to process data more efficiently, so organizations are able to adjust or make new plans with minimal resource costs.

Table 6.1 summarizes and details the representative empirical studies using OIPT. In the SCM literature, uncertainty is considered one of the most important factors influencing SCM. Given the need to collect, process and interpret large amounts of information, supply chain disruptions are identified as a major source of uncertainty. Due to the close internal relation between OIPT and SCM, researchers often apply OIPT as a theoretical lens to examine all sorts of phenomena such as sustainable supply chain management (SSCM) (Busse et al., 2017; Fabbe-Costes et al., 2014), internal and external SCI (Flynn et al., 2016; Srinivasan & Swink, 2015; Swink et al., 2007; Williams et al., 2013; Wong et al., 2011), the impact of information processing on the effectiveness of supply chain practices (Zhou & Benton, 2007), supply chain disruption risks (Bode & Macdonald, 2017; Bode et al., 2011; DuHadway et al., 2019), supply chain resilience (Dubey et al., 2021; Wong et al., 2015a) and etc. In recent studies, scholars have emphasized the importance of information aspects of SCRM from the perspective of OIPT. For example, the relationships between risk management practices, external integration, exogenous disruption risk, supply chain planning and operational performance have been scrutinized (Kauppi et al., 2016; Srinivasan & Swink, 2015). Furthermore, a recent study presented by Shou et al. (2018a) investigated the impact of SCRM on financial performance, operational efficiency and operational flexibility, as well as the moderating effect of supplier integration in the relationship between SCRM and operating performance.

6.2.2 SCRM and Operational Performance

Operational effectiveness, which denotes the differentiation performance in terms of quality, delivery and flexibility (Golini et al., 2016; Szász et al., 2016), can be achieved through rich buffering information and faster responsiveness provided by SCRM practices (Williams et al., 2013). Moreover, previous research also argues that higher information processing capability generated from SCRM practices increases the quick response and customization ability to meet customer requirements (Fan et al., 2017).

Table 6.1 Representative empirical studies of OIPT

Level of analysis	Study	Factors related to capacity	Factors related to needs	Research findings
Inter-organizational	Flynn and Flynn (1999)	Goal diversity; Reduction of sources of manufacturing environment complexity; Investment in manufacturing information systems; The use of lateral relations	Manufacturing environment complexity	At least one information-processing alternative, including self-contained tasks, lateral relations, and environmental management strategies for reducing manufacturing, supplier, and goal diversity moderates the negative relationship between environmental complexity and manufacturing performance
	Kim et al. (2005)	Electronic information transfer (EIT)	Demand uncertainty; Channel interdependence; Product characteristics	The monitoring component of EIT has a significant influence on demand uncertainty, and complexity-in-use is influenced by the coordination aspect of EIT. Both the coordination and monitoring aspects of EIT are significantly relevant to interdependence of partners in a supply channel
	Premkumar et al. (2005)	Information technology support	Environmental uncertainty; Partnership uncertainty	The interactive effect of information needs and capability has a significant effect on performance, supporting the fit theory

(continued)

6.2 Literature Review and Hypotheses Development

Table 6.1 (continued)

Level of analysis	Study	Factors related to capacity	Factors related to needs	Research findings
	Stock and Tatikonda (2008)	Inter-organizational interaction	Technology uncertainty	External technology integration will be most successful when the level of interaction between the source of the technology and recipient of the technology is appropriately matched, or fit, to the characteristics of the technology to be integrated
	Mani et al. (2010)	Governance structure; Relational processes; Relational technologies	Complexity of the task environment; Interdependencies of the task environment	The extent to which firms can accurately predict their information requirements of control and coordination and design information capabilities that are aligned with such information requirements is an important determinant of their ability to leverage business process outsourcing
	Cegielski et al. (2012)	Cloud computing	Environmental uncertainty; Task uncertainty; Inter-organizational uncertainty	Information processing requirements and information processing capability affect intention to adopt cloud computing. The decision to adopt cloud computing is based upon complex circumstances

(continued)

Table 6.1 (continued)

Level of analysis	Study	Factors related to capacity	Factors related to needs	Research findings
	Wang et al. (2013)	IT-Enabled planning and control; Normative contracts	Supply chain uncertainty	As buyers and suppliers utilize the IT and relational solutions, they induce relation-specific responses represented as suppliers' business process investments and modification flexibility, which in turn lead to positive buyers
	Wong et al. (2015b)	IT-enabled collaborative decision making; IT infrastructure in the relationships among inter-organizational information integration	-	Inter-organizational information integration is positively related to IT-enabled collaborative decision making, leading to customer service performance, when a high level of IT infrastructure development is present
	Wong et al. (2015a)	Supply chain information integration	Product complexity; Market complexity	Supply chain information integration facilitates greater performance improvements when it serves less complex products or is operated under a highly complex market environment

(continued)

6.2 Literature Review and Hypotheses Development

Table 6.1 (continued)

Level of analysis	Study	Factors related to capacity	Factors related to needs	Research findings
	Fan et al. (2016)	Supply chain risk information sharing; Supply chain risk information analysis	Product complexity Product customization; Technology turbulence; Market turbulence	Except product complexity, product-specific and environment-related uncertainty characteristics positively moderate the positive relationship between supply chain risk information capability and operational performance
	Shou et al. (2018a)	Supplier integration	Supply chain risk management	SCRM positively influences both operational efficiency and flexibility, and has an indirect effect on financial performance Supplier integration enhances the impact of SCRM on operational flexibility, but does not moderate the relationship between SCRM and operational efficiency
	Belhadi et al. (2021)	Artificial intelligence-driven innovation; Supply chain resilience	Supply chain dynamism	While AI has a direct impact on supply chain performance in the short-term, it is recommended to exploit its information processing capabilities to build supply chain resilience for long-lasting supply chain performance

(continued)

Table 6.1 (continued)

Level of analysis	Study	Factors related to capacity	Factors related to needs	Research findings
	Gu et al. (2021)	Information technology usage; Supply chain resilience	–	Only explorative use of IT with suppliers and customers can facilitate supplier and customer resilience, while exploitative use of IT demonstrates no significant effects. The balanced use of IT with customers shows a negative effect on customer resilience and complementary use shows a positive effect
	Stock et al. (2021)	Inter-organizational interaction; Knowledge sharing quantity	New product development project uncertainty; Knowledge sharing requirements	Knowledge sharing mediates the relationship between uncertainty and performance. Failure to align exchange of knowledge to needs is harmful in either direction. External design agents must share knowledge to exact requirements
	Yang et al. (2021)	Supply chain disruption orientation; Supply chain visibility	Supply chain disruption impact	The fit between information processing capacities and requirements enhances supply chain risk management capabilities, which, in turn, result in enhanced supply chain resilience

(continued)

6.2 Literature Review and Hypotheses Development

Table 6.1 (continued)

Level of analysis	Study	Factors related to capacity	Factors related to needs	Research findings
Intra-organizational	Gattiker and Goodhue (2004)	Enterprise resource planning (ERP)	Interdependence; Differentiation	High interdependence among organizational sub-units contributes to the positive ERP-related effects because of ERPs ability to coordinate activities and facilitate information flows. When differentiation among sub-units is high, organizations may incur ERP-related compromise or design costs
	Fairbank et al. (2006)	Managing the environment; Creating self-contained units; Creating vertical information systems; Creating lateral coordinating relationships	Organization's strategic orientation	While some information processing design choices are generally related to organizational performance, others should be matched to a specific strategic posture

(continued)

Table 6.1 (continued)

Level of analysis	Study	Factors related to capacity	Factors related to needs	Research findings
	Gattiker (2007)	Enterprise resource planning	Manufacturing–marketing interdependence	The greater ERP-enabled coordination between manufacturing and marketing, the greater the benefit of ERP to the plant The degree to which ERP-enabled manufacturing–marketing coordination improvements are realized depends on the amount of interdependence between manufacturing and marketing

(continued)

6.2 Literature Review and Hypotheses Development

Table 6.1 (continued)

Level of analysis	Study	Factors related to capacity	Factors related to needs	Research findings
	Gupta et al. (2019)	Information systems agility; Human resource performance management systems	-	Agile project management and agile software development is positively related to the diagnostic and interactive use of the performance measurement system. Both diagnostic and interactive use of the performance measurement system is positively related to the job satisfaction. Agile project management is positively related to job satisfaction. The relationship of agile team job satisfaction is negatively affected by the agile software development
Inter- and intra-organizational	Tatikonda and Montoya-Weiss (2001)	Process concurrency; Process formality; Process adaptability	Technological uncertainty; Market uncertainty; Environmental uncertainty	The relationships between organizational process factors and operational outcomes, and between operational outcomes and market outcomes, were found to be quite robust under varying conditions of technological, market, and environmental uncertainty

(continued)

Table 6.1 (continued)

Level of analysis	Study	Factors related to capacity	Factors related to needs	Research findings
	Srinivasan and Swink (2015)	Supply chain integration; Supply chain management systems; Planning comprehensiveness	-	The usage of supply chain management systems enables organizations to better utilize the information they gain from external integration efforts, thus improving the comprehensiveness of their supply chain planning capabilities The use of supply chain management systems appears to be a partial substitute for internal integration as a driver of planning comprehensiveness

(continued)

Table 6.1 (continued)

Level of analysis	Study	Factors related to capacity	Factors related to needs	Research findings
	Srinivasan and Swink (2018)	Analytics capability; Supply chain visibility; Organizational flexibility	Market volatility	Analytics capability is more strongly associated with operational performance when supply chain organizations also possess organizational flexibility needed to act upon analytics-generated insights quickly and efficiently Analytics capability and organizational flexibility are more valuable as complementary capabilities for firms who operate in volatile markets, rather than in stable ones
	Zhu et al. (2018)	Supply chain analytics (SCA)	Supply uncertainty	Analytics capability in support of planning functions indirectly affects organizational supply chain transparency (OSCT) via SCA capabilities in source, make, and deliver functions SCA capabilities in source, make, and deliver positively influence OSCT Supply uncertainty moderates the relationship between SCA capabilities in make and OSCT

(continued)

Table 6.1 (continued)

Level of analysis	Study	Factors related to capacity	Factors related to needs	Research findings
	Dubey et al. (2021)	Data analytics capability; Organizational flexibility	-	Data analytic capability is a means by which visibility improves supply chain resilience and leads to competitive advantage
	Li et al. (2020)	Digital technologies	Environmental dynamism	Digital supply chain platforms mediate the effects of digital technologies on both economic and environmental performance The mediating effects are enhanced under a high degree of environmental dynamism
	Wong et al. (2020)	Supply chain resilience	Supply side disruptions; Infrastructure disruptions; Catastrophic disruptions	Supply chain resilience positively contributes to risk management, market, and financial performance Supply chain resilience pays under high levels of supply, infrastructure, and catastrophic disruptions

(continued)

6.2 Literature Review and Hypotheses Development

Table 6.1 (continued)

Level of analysis	Study	Factors related to capacity	Factors related to needs	Research findings
	Yu et al. (2021)	Big data analytics capability; Hospital supply chain integration	-	Big data analytics capability has a significant impact on three dimensions of hospital SCI: inter-functional integration, hospital-patient integration, and hospital-supplier integration. Hospital-patient integration and hospital-supplier integration fully mediate the relationship between inter-functional integration and operational flexibility

Regarding operational efficiency, which is defined as the performance of cost and lead time (Golini et al., 2016; Szász et al., 2016), the mainstream literature on SCRM argues that a firm can gain efficiency improvement through acquiring and processing rich and accurate operations and supply chain information from SCRM practices (Kauppi et al., 2016; Wiengarten et al., 2016). In addition, the risk information processing capability (e.g., risk information sharing, risk analysis and assessment) created by SCRM practices contributes to operational efficiency because firms can better deal with operational information, reduce the resources waste and time loss, and prevent operations risks (Lam et al., 2016; Narasimhan & Talluri, 2009). However, some studies argue that SCRM practices may decrease a firm's operational efficiency in that additional investments are required in terms of excess inventories, backup suppliers, preventive maintenance and extra capabilities (Shou et al., 2018a). Based on the above review of the performance effects of SCRM practices, we propose the following hypothesis:

H1 SCRM practices are positively associated with a manufacturing firm's (a) operational effectiveness and (b) operational efficiency.

6.2.3 The Moderating Effect of IAD

In the era of globalization, manufacturing MNCs are increasingly adopting GMNs to pursue lower cost and exploit location-specific resources and advantages (Leuschner et al., 2013; Manuj & Mentzer, 2008a). Therefore, IAD has attracted much attention in the extant SCM literature. IAD indicates the geographical distance and spatial complexity of manufacturing MNCs (Goerzen & Beamish, 2003; Zaheer & Hernandez, 2011) and may weaken the accuracy and timeliness of the production and managerial information (Bode & Wagner, 2015). Consequently, it is conjectured that IAD may influence the efficacy of SCRM practices on operational performance (Goerzen & Beamish, 2003; Zaheer & Hernandez, 2011).

Regarding operational effectiveness, the buffering information provided by SCRM practices such as production and alternative transportation modes may be obfuscated and delayed by broader IAD (Bode & Wagner, 2015; Williams et al., 2013). Such inaccurate and delayed information may hinder the effective processing of operational information and further impair timely delivery and manufacturing flexibility (Stock et al., 2000). Second, broader IAD also increases the administrative barrier to efficiently gather and process the risk information provided by SCRM practices and weakens the accuracy and timeliness of risk information from SCRM practices (Goerzen & Beamish, 2003; Madhavan et al., 2004), which may weaken the positive impact of SCRM practices on operational effectiveness.

Similarly, broader IAD may also lessen the efficacy of SCRM practices on operational efficiency. First, with broader IAD, the efficacy of firm's SCRM practices on operational efficiency may decrease because IAD impairs the accuracy, transparency, and timeliness of operational information provided by SCRM practices, which may

6.2 Literature Review and Hypotheses Development

trigger firms to invest more in practices such as backup suppliers and preventive maintenance. Second, the associated demand and supply uncertainty induced by broader IAD requires extra buffer strategies, which will increase production cost and lead time (Bode & Wagner, 2015; Cheng et al., 2015; Habermann et al., 2015). Furthermore, the spatial complexity of broader IAD undermines the firm's risk information processing capability and further decrease the effectiveness of SCRM practices in promoting operational efficiency (Goerzen & Beamish, 2003, 2005). Consequently, the efficacy of SCRM practices on operational efficiency may decrease under broader IAD. Therefore, we hypothesize that:

H2 The broader the IAD, the weaker the efficacy of SCRM practices on a manufacturing MNC's (a) operational effectiveness and (b) operational efficiency.

6.2.4 The Counteracting Effect of SCI

Extant literature has extensively studied the information-facilitating role of SCI (Flynn et al., 2010; Schoenherr & Swink, 2012) and identified different dimensions of SCI such as internal integration, manufacturing network integration (MNI), and external integration (EI) (Cheng & Farooq, 2018; Kim & Schoenherr, 2018; Szász et al., 2016). By sharing in-time production and supply chain information, all three types of SCI enable the higher visibility of production procedures and better collaboration across functions, plants in manufacturing networks, and among firms in the supply chain networks (Williams et al., 2013; Wong et al., 2011). The information-intensive attribute of the three types of SCI offers potential to mitigate the attenuating effect of IAD on the relationship between SCRM practices and operational performance through providing rich and high-quality operations and supply chains information.

First, internal integration incorporates the multiple functions of a firm and facilitates internal operational information acquisition from different functions and departments (Schoenherr & Swink, 2012), which compensates the inaccurate and delayed information induced by broader IAD and reduces the firm's information processing requirements. Second, based on the accurate information obtained from internal integration, firms can offset the obfuscation of information caused by IAD and develop more appropriate buffering strategies such as relatively accurate excess inventories and extra capabilities, and limited number of backup suppliers (Kauppi et al., 2016). Finally, firms with higher levels of internal integration tends to have advanced information systems and higher managerial ability, which increases the information processing capabilities and enables the mitigation of the negative effect of IAD on SCRM practices (Liu et al., 2018; Srinivasan & Swink, 2015).

For firms with manufacturing networks, better MNI enables the gathering and assimilation of operational information across multiple plants (Szász et al., 2016). Thus, these plants are more capable of identifying and assessing risks and taking SCRM actions proactively even if they have broader IAD. Moreover, supported by advanced technology-based mechanisms, MNI promotes information sharing and transfer among multiple plants as well as supply chain members (Szász et al., 2016; Williams et al., 2013). Such extensive and efficient information sharing within a manufacturing network offsets the negative effect of IAD and contributes to better response to supply chain uncertainties such as demand fluctuations.

Prior studies have disclosed the significant role of supply chain relationships in information sharing and processing (Chen et al., 2013; Kauppi et al., 2016; Li et al., 2015; Wiengarten et al., 2016). First, external integration (including supplier integration and customer integration) ensures the external information accessibility and accuracy (Tang & Nurmaya Musa, 2011). The rich and accurate real-time upstream and downstream information acquired from the collaboration and coordination with supply chain partners helps reduce the obstruction introduced by IAD (Bode et al., 2011; Chen et al., 2013; Zeng & Yen, 2017). Second, strategic collaboration with key external supply chain partners enhances focal firms' ability to quickly respond to unpredictable changes at both supply and demand sides (Golini & Gualandris, 2018; Klassen & Vachon, 2003; Liu et al., 2016), which also mitigates the delay caused by broader IAD. Therefore, we propose:

H3 The negative moderating effect of IAD on the relationship between SCRM practices and (a) operational effectiveness and (b) operational efficiency will be weakened as the level of internal integration increases.

H4 The negative moderating effect of IAD on the relationship between SCRM practices and (a) operational effectiveness and (b) operational efficiency will be weakened as the level of manufacturing network integration increases.

H5 The negative moderating effect of IAD on the relationship between SCRM practices and (a) operational effectiveness and (b) operational efficiency will be weakened as the level of external integration increases.

Figure 6.1 depicts the conceptual model of this study.

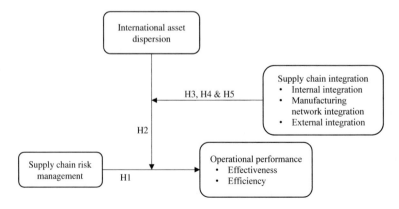

Fig. 6.1 Conceptual model

6.3 Method

6.3.1 Data

The empirical analysis of this study is based on the data of the sixth round International Manufacturing Strategy Survey (IMSS). A detailed description of the IMSS dataset is reported in Chap. 1 of this book (Shou et al., 2022). Since this study investigates the effects of IAD and MNI, 307 responses from stand-alone plants were excluded. 246 responses were further dropped owing to missing data. Finally, 378 responses were included in the final sample, in which 105 responses had their plants located in a single country, 64 responses had multiple plants located in one continent, and 209 responses had plants located globally.

6.3.2 Measures

All the measures were adapted from the extant literature. The detailed survey questions of the constructs are presented in the Appendix of this book (Shou et al., 2022).

Supply chain risk management. SCRM is measured by the implementation of the prevention, detection, respond to, and recovery from operations and supply chain risks (Chaudhuri et al., 2018; Jüttner et al., 2003; Kauppi et al., 2016). The respondents are requested to "indicate the effort put in the last three years into implementing action programs" related to SCRM, with the scale ranging from 1 (low) to 5 (high).

International asset dispersion. This study measures IAD by different types of international manufacturing networks including multiple plants in one country, multiple plants in one continent, or multiple plants in multiple continents (Wiengarten & Longoni, 2018). Although some previous studies have operationalized IAD by the numbers of foreign countries that the manufacturing MNCs operate, some other studies also provide support for our measure of IAD in terms of assets (i.e. plants and facilities) diversification across the world (Fang et al., 2007, 2013).

Supply chain integration. Following prior extensive research on the types of SCI and its measurement, three types of SCI are measured by five-point Likert scales. Internal integration (II) was measured by four items which indicates the level of adoption of the information sharing and joint decision-making across different departments within a firm (Flynn et al., 2010). Manufacturing network integration (MNI) was measured through five items with regard to the implementation of information sharing, joint decision making, joint innovation, technological support to communication, and network performance management system at the inter-plant level (Cheng et al., 2016). External integration (EI), including both supplier integration and customer integration, was measured by eight items which reflect the level of adoption by focal firms in terms of the information sharing, collaboration, joint decision making and system coupling with suppliers and customers (Frohlich & Westbrook, 2001).

Operational effectiveness and efficiency. This study focuses on the operational performance in terms of the improvement of effectiveness and efficiency, which is consistent with the classic strategic management literature (Porter, 1985). These two types of operational performance are also adopted by previous SCM literature (Demeter et al., 2016; Golini et al., 2016; Szász et al., 2016) as well as international business (e.g., Luo, 2007; Zhou & Li, 2008). Specifically, operational effectiveness focuses on the improvement in quality, delivery and flexibility, which were measured by six items as one item for flexibility (F3) was dropped. Operational efficiency focuses on the improvement of cost and time, which were measured by four items (Golini et al., 2016; Szász et al., 2016). Both types of operational performance are measured by questions that "How does your current performance compare with that of your main competitor?" with the scale ranging from 1 (much lower) to 5 (much higher).

Control variables. This study considers several control variables in order to mitigate the potential omitted variables bias (Angrist & Pischke, 2009). First, we control for plant size because larger plants tend to have more resources and perform better in terms of effectiveness and efficiency. Plant size was measured by the natural logarithm of the number of employees. Second, firms in different industries may experience different levels of performance because of the industry characteristics such as competition intensity (Liu et al., 2016). Dummy variables for industries were created to account for systematic differences across industries (Lu & Shang, 2017). We also control for product complexity since it may have significant influence on firms' operational performance (Thomé & Sousa, 2016). Product complexity was measured by items related to the level of product design integration, the numbers of parts or materials, and the number of steps required (Shou et al., 2018b). The adoption of digital manufacturing technology (DMT), which was measured by three items, is also included since it may influence operational performance (Thomé & Sousa, 2016). In addition, the demand and supply uncertainty in the supply chain has been proved to affect firm performance (Tang, 2006; Wiengarten & Longoni, 2018). Four items that are associated with the demand uncertainty and supply uncertainty were applied to assess supply chain uncertainty (Wiengarten & Longoni, 2018). Similarly, supply chain disruption risk, which denotes the probability of occurrence of supply chain disruption, was controlled for and measured by three items (Chaudhuri et al., 2018). Finally, we controlled for production type since it is a key characteristic of plants and can influence operational performance (Safizadeh et al., 2000). Production type was measured by the percentage of mass production in total production (Szász et al., 2016).

6.3.3 Reliability and Validity

In this study, IBM SPSS 22.0 was used to analyze the reliability of the constructs. Confirmatory factor analysis (CFA) was conducted by Mplus 7.4 to test the reliability and validity of the construct measurements. The results are presented in Table 6.2. The Cronbach's alpha of each construct was greater than 0.6 (Flynn et al., 1990). The composite reliability (CR) value of each construct exceeded the criterion of 0.70 (Nunnally, 1978) except for product complexity (0.697), which is very close to the threshold and acceptable (Hair et al., 2010). Therefore, the results confirm the internal consistent reliability of the measurements.

We assessed the validity of our study in terms of content, convergent and discriminant validity. Content validity was guaranteed since the sixth round IMSS was developed by a team of senior researchers and extracted from solid literature. The CFA results showed an acceptable fit of the measurement model ($\chi^2/df = 2.294$; RMSEA $= 0.059$; CFI $= 0.872$; TLI $= 0.859$; SRMR $= 0.051$). All factor loadings were significant at the level of 0.001 and were greater than the recommended value of 0.50 except for PC1 (0.423), which is still acceptable (Atuahene-Gima & Li, 2004). Besides, the standardized coefficients for all items were greater than twice their standard errors. The CFA results indicated the convergent validity of the measures (Bollen, 1989). The estimates of average variance extracted (AVE) for all the constructs were greater than 0.40 and all AVE estimates were less than the corresponding CR values (Hair et al., 2010). The above-mentioned results indicate the convergent validity of our study. The square root of AVE value for each construct is larger than any corresponding correlation coefficient (see Table 6.3), which provides evidence of discriminant validity (Fornell & Larcker, 1981).

6.4 Results

In this study, hierarchical regression analysis was used to test the proposed hypotheses. Since we applied two-way and three-way interaction terms to test the moderating effects of IAD and SCI, we mean-centered SCRM, IAD, and three types of SCI to mitigate the potential threat of multicollinearity before generating the two-way and three-way interaction terms (Aiken & West, 1991). In addition, we also calculated the variance inflation factors (VIFs) to check multicollinearity. All VIFs are well below cut-off value 10 (Su et al., 2013), which suggests that multicollinearity is not a serious concern.

All variables and the interaction terms were introduced to the hierarchical models for the two dependent variables (i.e., operational effectiveness and operational efficiency) following a stepwise approach. First, we included all the control variables in the base models (Model 1 and Model 6). Then, we included the independent variable (Model 2 and Model 7) and all the moderators (Model 3 and Model 8) to examine the main effect. Third, the two-way interaction terms between SCRM and

Table 6.2 CFA analysis results

Construct	Item	Factor loading	S.E	t-Value	R^2
Supply chain risk management (Cronbach's α = 0.862; CR = 0.863; AVE = 0.613)	SCRM1	0.740	0.028	26.639	0.548
	SCRM2	0.831	0.022	38.164	0.691
	SCRM3	0.795	0.024	32.962	0.632
	SCRM4	0.762	0.026	29.080	0.581
Internal integration (Cronbach's α = 0.886; CR = 0.887; AVE = 0.663)	II1	0.772	0.025	30.544	0.596
	II2	0.820	0.022	36.585	0.672
	II3	0.851	0.020	43.204	0.724
	II4	0.811	0.023	35.928	0.658
Manufacturing network integration (Cronbach's α = 0.875; CR = 0.876; AVE = 0.586)	MNI1	0.795	0.023	34.008	0.632
	MNI2	0.769	0.025	30.461	0.592
	MNI3	0.728	0.028	25.920	0.530
	MNI4	0.752	0.026	28.556	0.566
	MNI5	0.782	0.024	32.233	0.611
External integration (Cronbach's α = 0.902; CR = 0.902; AVE = 0.536)	SI1	0.687	0.031	22.463	0.472
	SI2	0.753	0.026	28.666	0.567
	SI3	0.765	0.025	31.106	0.586
	SI4	0.718	0.028	25.875	0.516
	CI1	0.773	0.025	31.312	0.598
	CI2	0.748	0.027	28.185	0.560
	CI3	0.701	0.029	23.824	0.491
	CI4	0.708	0.029	24.578	0.502
Operational effectiveness (Cronbach's α = 0.803; CR = 0.804; AVE = 0.411)	Q1	0.598	0.041	14.720	0.357
	Q2	0.601	0.041	14.733	0.361
	D1	0.743	0.033	22.529	0.552
	D2	0.766	0.031	24.443	0.587
	F1	0.568	0.042	13.614	0.323
	F2	0.532	0.044	12.121	0.283
Operational efficiency (Cronbach's α = 0.764; CR = 0.766; AVE = 0.451)	C1	0.694	0.040	17.252	0.482
	C2	0.649	0.042	15.499	0.421
	T1	0.673	0.040	16.724	0.454
	T2	0.668	0.040	16.573	0.446
Product complexity (Cronbach's α = 0.661; CR = 0.697; AVE = 0.454)	PDC1	0.423	0.052	8.143	0.179
	PDC2	0.627	0.047	13.281	0.394
	PDC3	0.889	0.053	16.833	0.791
Digital manufacturing technology (Cronbach's α = 0.782; CR = 0.789; AVE = 0.557)	DMT1	0.662	0.036	18.596	0.439
	DMT2	0.834	0.029	28.973	0.696

(continued)

6.4 Results

Table 6.2 (continued)

Construct	Item	Factor loading	S.E	t-Value	R^2
	DMT3	0.732	0.033	22.284	0.536
Supply chain disruption risk (Cronbach's $\alpha = 0.861$; CR $= 0.866$; AVE $= 0.684$)	SCDR1	0.743	0.028	26.878	0.552
	SCDR2	0.918	0.020	45.450	0.843
	SCDR3	0.810	0.024	33.570	0.656
Supply chain uncertainty (Cronbach's $\alpha = 0.887$; CR $= 0.889$; AVE $= 0.667$)	SCU1	0.800	0.024	33.217	0.640
	SCU2	0.831	0.022	37.050	0.690
	SCU3	0.780	0.025	30.682	0.609
	SCU4	0.853	0.021	40.696	0.728

IAD were included to test the moderating effect of IAD (Model 4 and Model 9). Finally, the three-way interaction terms between SCRM, IAD and three types of SCI were included (Model 5 and Model 10). Table 6.4 presents the hierarchical regression results of the models.

H1 proposed that SCRM practices are positively associated with firms' operational effectiveness and efficiency. The results in Table 6.4 confirm that SCRM practices are positively associated with operational effectiveness ($\beta = 0.094, p = 0.032$). However, SCRM practices are not significantly associated with operational efficiency ($\beta = -0.053, p = 0.258$). Hence, H1a is supported whereas H1b is not.

We proposed the negative moderating effect of IAD on the relationships between SCRM practices and operational effectiveness and efficiency in H2a and H2b respectively. As shown in Table 6.4, H2a is supported ($\beta = -0.089, p = 0.029$) whereas H2b is not ($\beta = -0.040, p = 0.360$). The results suggest that the efficacy of SCRM practices in manufacturing MNCs is attenuated by broader IAD in terms of operational effectiveness. However, IAD does not show significant moderating effect on the relationship between SCRM practices and operational efficiency.

Regarding the three-way interaction terms, H3a, H4a and H5a proposed that three types of SCI (i.e., II, MNI and EI) can counteract the negative moderating effect of IAD on the relationship between SCRM practices and operational effectiveness. According to the regression results, H5a is confirmed ($\beta = 0.127, p = 0.062$) whereas H3a and H4a are not supported. Hence, EI is the only one that has a significant moderating effect. Similarly, H3b, H4b and H5b proposed that three types of SCI can mitigate the attenuating effect of IAD on SCRM–operational efficiency relationship, which are not supported by the results.

Table 6.3 Descriptive statistics and correlation matrix

Constructs	Mean	SD	Min	Max	(1)	(2)	(3)	(4)	(5)
(1) Supply chain risk management	3.35	0.81	1	5	0.783				
(2) International asset dispersion	3.28	0.87	2	4	0.088	-			
(3) Internal integration	3.48	0.88	1	5	0.524**	0.054	0.814		
(4) Manufacturing network integration	3.26	0.87	1	5	0.520**	0.126*	0.544**	0.766	
(5) External integration	3.10	0.84	1	5	0.575**	0.094	0.615**	0.590**	0.732
(6) Operational effectiveness	3.52	0.57	1	5	0.337**	0.061	0.247**	0.311**	0.341**
(7) Operational efficiency	3.12	0.57	1	5	0.141**	0.005	0.172**	0.280**	0.238**
(8) Plant size	6.41	1.77	1.10	11.84	0.130*	0.188**	0.074	0.122*	0.202**
(9) Product complexity	3.61	0.86	1	5	0.188**	0.042	0.232**	0.182**	0.263**
(10) Digital manufacturing technology	3.00	1.01	1	5	0.381**	−0.001	0.280**	0.385**	0.461**
(11) Supply chain disruption risk	2.58	1.04	1	5	0.086	−0.107*	0.021	0.113*	0.114*
(12) Supply chain uncertainty	2.75	0.96	1	5	0.020	0.010	0.055	0.050	0.153*
(13) Production type	0.25	0.34	0	1	0.137*	−0.053	0.111*	0.147**	0.188*

(continued)

Table 6.3 (continued)

Constructs	Mean	SD	Min	Max	(1)	(2)	(3)	(4)	(5)
	(6)	(7)	(8)	(9)	(10)	(11)	(12)	(13)	
(6) Operational effectiveness	0.641								
(7) Operational efficiency	0.401**	0.672							
(8) Plant size	0.084	0.092	-						
(9) Product complexity	0.161**	0.028	0.142**	0.674					
(10) Digital manufacturing technology	0.358**	0.193**	0.134**	0.271**	0.746				
(11) Supply chain disruption risk	0.076	0.103*	0.037	0.084	0.098	0.827			
(12) Supply chain uncertainty	0.027	0.099	−0.055	0.047	0.026	0.258**	0.817		
(13) Production type	0.182**	0.097	0.141**	−0.042	0.192**	−0.004	−0.096	-	

Note The values of the diagonal elements are the squared-root of AVE
*Correlation is significant at the 0.05 level
**Correlation is significant at the 0.01 level

Table 6.4 Results of OLS regression analyses

	Operational effectiveness					Operational efficiency				
	Model 1	Model 2	Model 3	Model 4	Model 5	Model 6	Model 7	Model 8	Model 9	Model 10
Control variables										
Industry dummies	Yes	Yes	Yes	Yes	Yes	Yes	Yes	Yes	Yes	Yes
Plant size	0.001	−0.003	−0.010	−0.007	−0.009	0.015	0.014	0.010	0.012	0.008
Product complexity	0.043	0.031	0.023	0.029	0.020	−0.039	−0.043	−0.055	−0.043	-0.055
Digital manufacturing technology	0.177***	0.138***	0.117***	0.136***	0.117***	0.090***	0.077**	0.045	0.077**	0.038
Supply chain disruption risk	0.020	0.014	0.016	0.014	0.014	0.038	0.036	0.033	0.035	0.033
Supply chain uncertainty	0.034	0.034	0.023	0.026	0.018	0.056*	0.056*	0.046	0.052	0.042
Production type	0.222***	0.198***	0.184**	0.202***	0.170**	0.089	0.082	0.053	0.082	0.052
Independent variable										
Supply chain risk management (SCRM)		0.150***	0.096**	0.149***	0.094**		0.045	−0.047	0.045	-0.053
Moderating variables										
International asset dispersion (IAD)			0.032	0.041	0.030			−0.004	0.013	0.005
Internal integration (II)			−0.003		0.001			0.015		0.023
Manufacturing network integration (MNI)			0.056		0.059			0.152***		0.145***
External integration (EI)			0.060		0.054			0.035		0.045

(continued)

Table 6.4 (continued)

	Operational effectiveness					Operational efficiency				
	Model 1	Model 2	Model 3	Model 4	Model 5	Model 6	Model 7	Model 8	Model 9	Model 10
Interaction terms										
$SCRM \times IAD$				−0.080**	−0.089**				−0.045	-0.040
$SCRM \times IAD \times II$					−0.073					-0.095
$SCRM \times IAD \times MNI$					−0.051					-0.006
$SCRM \times IAD \times EI$					0.127*					0.078
Number of observations	378	378	378	378	378	378	378	378	378	378
Adjusted R square	0.166	0.203	0.210	0.211	0.218	0.049	0.050	0.085	0.048	0.085
F value	7.84***	9.00***	7.26***	8.19***	6.26***	2.76***	2.64**	3.19***	2.34**	2.76***

$*p < 0.1$; $**p < 0.05$; $***p < 0.01$

6.5 Discussion

6.5.1 Findings

While SCRM practices are generally believed to benefit firms' operational performance, the relationship between SCRM practices and operational performance is not clear due to the mixed evidence (e.g., Kauppi et al., 2016; Manuj et al., 2014; Narasimhan & Talluri, 2009; Premkumar et al., 2005) This study generates insights into the effects of SCRM practices on operational effectiveness and efficiency separately, as well as the moderating effects of IAD and three types of SCI in the relationship between SCRM practices and operational performance. In consistent with previous research, our results confirm the positive effect of SCRM practices on operational effectiveness (Kauppi et al., 2016; Wiengarten et al., 2016). Furthermore, the results also reveal that there is no significant relationship between SCRM practices and operational efficiency. One possible explanation for the rejection is that extra investment in buffering strategies such as excess inventories, backup suppliers, preventive maintenance and extra capabilities that SCRM practices requires (Shou et al., 2018a) may lead to cost increase and thus compromise operational efficiency.

Regarding H2, we find that IAD significantly weakens the efficacy of SCRM practices on operational effectiveness while it shows no significant moderating effect on the SCRM–operational efficiency relationship. In H3, H4, and H5, we further propose that three types of SCI can mitigate the negative effect of broader IAD. While it is confirmed that EI can counteract the attenuating effect of IAD on the relationship between SCRM practices and operational effectiveness, the counteracting effects of II and MNI are not significant, which indicates that boundary-spanning information sharing is more important than information sharing within a single manufacturing MNC. In other words, in contrast with the high level heterogeneity and low level transparency of information from not-well-integrated external partners (Paulraj et al., 2008; Zhou & Benton, 2007), better EI can counteract the negative effect of IAD with high-quality timely information from supply chain partners. Internal integration and MNI may help accelerate the information flow within a manufacturing MNC but do not contribute to better access of external information related to supply chain risks, thereby failing to provide counteracting effects. Consequently, EI plays a more significant role in strengthening the efficacy of SCRM practices in manufacturing MNCs with broader IAD.

6.5.2 Theoretical Implications

The theoretical contributions of this study are twofold. First, this study contributes to the SCM literature by explicitly investigating the distinctive effects of SCRM practices on operational effectiveness and efficiency. The results indicate that SCRM practices are beneficial to operational effectiveness; however, the effect on operational efficiency is not significant. In other words, SCRM practices are more important for manufacturing firms pursuing a differentiation strategy than those with a cost leadership strategy. The empirical evidence provides explanations for the mixed findings on the efficacy of SCRM practices in the extant literature.

Second, this study also extends the research scope of SCM by linking an important construct in international business research, i.e., IAD, to SCRM and SCI of manufacturing MNCs. This study reveals the attenuating effect of IAD and the counteracting effect of EI on the efficacy of SCRM practices of manufacturing MNCs. Considering the significant moderating effect of IAD, it is essential to investigate the SCM of manufacturing MNCs from an international business perspective, which provides a valuable future research direction. In addition, this study also contributes to the SCI literature. Previous research has investigated different types of SCI extensively. Nevertheless, this study indicates that compared to internal integration and MNI, EI is more important in the successful implementation of SCRM practices owing to its boundary-spanning role in information sharing, which compliments the extant literature.

6.6 Conclusions

This study extends the SCRM research by examining the moderating effects of IAD and three types of SCI on the efficacy of SCRM practices. Our results confirm that SCRM practices are beneficial to firms' operational effectiveness. Our results also suggest that although IAD negatively moderates the relationship between SCRM practices and operational effectiveness, manufacturing MNCs can leverage EI to mitigate the attenuating effect of IAD.

This study has a number of limitations that constrain the generalizability and interpretation of its findings. First, this study utilizes cross-sectional survey data, which lacks enough evidence to conclude the causal effects in the proposed hypotheses. Future research may replicate this research using longitudinal studies to confirm the temporal nature of those relationships. Second, this study uses a categorical variable to measure IAD since the specific number of countries that a manufacturing MNC operates is not available in the dataset. Future research can adopt more precise measurement for IAD. Third, this study conceptualized SCRM practices as an aggregate construct. Future research should further investigate the different dimensions of SCRM and their individual impact on operational performance.

Acknowledgements This work was supported by the National Natural Science Foundation of China under Grant Numbers 71472166 and 71821002; and the Humanities and Social Sciences Faculty Development Program of Zhejiang University.

References

Aiken, L. S., & West, S. G. (1991). *Multiple regression: testing and interpreting interactions.* Sage.

Angrist, J. D., & Pischke, J.-S. (2009). *Mostly harmless econometrics: An empiricist's companion.* Princeton University Press.

Atuahene-Gima, K., & Li, H. (2004). Strategic decision comprehensiveness and new product development outcomes in new technology ventures. *Academy of Management Journal, 47*(4), 583–597.

Belhadi, A., Mani, V., Kamble, S. S., Khan, S. A. R., & Verma, S. (2021). Artificial intelligence-driven innovation for enhancing supply chain resilience and performance under the effect of supply chain dynamism: An empirical investigation. *Annals of Operations Research.* https://doi.org/10.1007/s10479-021-03956-x

Bode, C., & Macdonald, J. R. (2017). Stages of supply chain disruption response: Direct, constraining, and mediating factors for impact mitigation. *Decision Sciences, 48*(5), 836–874.

Bode, C., & Wagner, S. M. (2015). Structural drivers of upstream supply chain complexity and the frequency of supply chain disruptions. *Journal of Operations Management, 36*(1), 215–228.

Bode, C., Wagner, S. M., Petersen, K. J., & Ellram, L. M. (2011). Understanding responses to supply chain disruptions: Insights from information processing and resource dependence perspectives. *Academy of Management Journal, 54*(4), 833–856.

Bollen, K. A. (1989). A new incremental fit index for general structural equation models. *Sociological Methods & Research, 17*(3), 303–316.

Braunscheidel, M. J., & Suresh, N. C. (2009). The organizational antecedents of a firm's supply chain agility for risk mitigation and response. *Journal of Operations Management, 27*(2), 119–140.

Brown, D.A. (2015). *Intel corporation managing risk end-to-end in Intel's supply chain.* Available at https://csrc.nist.gov/CSRC/media/Projects/Supply-Chain-Risk-Management/documents/case_studies/USRP_NIST_Intel_100715.pdf. Accessed 25 Sept 2021.

Busse, C., Meinlschmidt, J., & Foerstl, K. (2017). Managing information processing needs in global supply chains: A prerequisite to sustainable supply chain management. *Journal of Supply Chain Management, 53*(1), 87–113.

Cegielski, C. G., Allison Jones-Farmer, L., Wu, Y., & Hazen, B. T. (2012). Adoption of cloud computing technologies in supply chains. *International Journal of Logistics Management, 23*(2), 184–211.

Chaudhuri, A., Boer, H., & Taran, Y. (2018). Supply chain integration, risk management and manufacturing flexibility. *International Journal of Operations & Production Management, 38*(3), 690–712.

Chen, J., Sohal, A. S., & Prajogo, D. I. (2013). Supply chain operational risk mitigation: A collaborative approach. *International Journal of Production Research, 51*(7), 2186–2199.

Cheng, Y., Chaudhuri, A., & Farooq, S. (2016). Interplant coordination, supply chain integration, and operational performance of a plant in a manufacturing network: A mediation analysis. *Supply Chain Management: an International Journal, 21*(5), 550–568.

Cheng, Y., & Farooq, S. (2018). The role of plants in manufacturing networks: A revisit and extension. *International Journal of Production Economics, 206*, 15–32.

Cheng, Y., Farooq, S., & Johansen, J. (2015). International manufacturing network: Past, present, and future. *International Journal of Operations & Production Management, 35*(3), 392–429.

Colicchia, C., & Strozzi, F. (2012). Supply chain risk management: A new methodology for a systematic literature review. *Supply Chain Management: an International Journal, 17*(4), 403–418.

Daft, R. L., & Lengel, R. H. (1986). Organizational information requirements, media richness and structural design. *Management Science, 32*(5), 554–571.

Demeter, K., Szász, L., & Rácz, B.-G. (2016). The impact of subsidiaries' internal and external integration on operational performance. *International Journal of Production Economics, 182*, 73–85.

References 123

Dubey, R., Gunasekaran, A., Childe, S. J., Fosso Wamba, S., Roubaud, D., & Foropon, C. (2021). Empirical investigation of data analytics capability and organizational flexibility as complements to supply chain resilience. *International Journal of Production Research, 59*(1), 110–128.

DuHadway, S., Carnovale, S., & Hazen, B. (2019). Understanding risk management for intentional supply chain disruptions: Risk detection, risk mitigation, and risk recovery. *Annals of Operations Research, 283*(1–2), 179–198.

Fabbe-Costes, N., Roussat, C., Taylor, M., & Taylor, A. (2014). Sustainable supply chains: A framework for environmental scanning practices. *International Journal of Operations & Production Management, 34*(5), 664–694.

Fairbank, J. F., Labianca, G., & "Joe", Steensma, H.K. and Metters, R. (2006). Information processing design choices, strategy, and risk management performance. *Journal of Management Information Systems, 23*(1), 293–319.

Fan, H., Cheng, T. C. E., Li, G., & Lee, P. K. C. (2016). The effectiveness of supply chain risk information processing capability: An information processing perspective. *IEEE Transactions on Engineering Management, 63*(4), 414–425.

Fan, H., Li, G., Sun, H., & Cheng, T. C. E. (2017). An information processing perspective on supply chain risk management: Antecedents, mechanism, and consequences. *International Journal of Production Economics, 185*(3), 63–75.

Fang, Y., Wade, M., Delios, A., & Beamish, P. W. (2007). International diversification, subsidiary performance, and the mobility of knowledge resources. *Strategic Management Journal, 28*(10), 1053–1064.

Fang, Y., Wade, M., Delios, A., & Beamish, P. W. (2013). An exploration of multinational enterprise knowledge resources and foreign subsidiary performance. *Journal of World Business, 48*(1), 30–38.

Flynn, B. B., & Flynn, E. J. (1999). Information-processing alternatives for coping with manufacturing environment complexity. *Decision Sciences, 30*(4), 1021–1052.

Flynn, B. B., Huo, B., & Zhao, X. (2010). The impact of supply chain integration on performance: A contingency and configuration approach. *Journal of Operations Management, 28*(1), 58–71.

Flynn, B. B., Koufteros, X., & Lu, G. (2016). On theory in supply chain uncertainty and its implications for supply chain integration. *Journal of Supply Chain Management, 52*(3), 3–27.

Flynn, B. B., Sakakibara, S., Schroeder, R. G., Bates, K. A., & Flynn, E. J. (1990). Empirical research methods in operations management. *Journal of Operations Management, 9*(2), 250–284.

Fornell, C., & Larcker, D. F. (1981). Evaluating structural equation models with unobservable variables and measurement error. *Journal of Marketing Research, 18*(1), 39–50.

Frohlich, M. T., & Westbrook, R. (2001). Arcs of integration: An international study of supply chain strategies. *Journal of Operations Management, 19*(2), 185–200.

Galbraith, J. R. (1974). Organization design: An information processing view. *Interfaces, 4*(3), 28–36.

Galbraith, J. R. (1974). *Designing complex organizations* (Vol. 27). Addison-Wesley.

Ganbold, O., Matsui, Y., & Rotaru, K. (2021). Effect of information technology-enabled supply chain integration on firm's operational performance. *Journal of Enterprise Information Management, 34*(3), 948–989.

Gattiker, T. F. (2007). Enterprise resource planning (ERP) systems and the manufacturing–marketing interface: An information-processing theory view. *International Journal of Production Research, 45*(13), 2895–2917.

Gattiker, T. F., & Goodhue, D. L. (2004). Understanding the local-level costs and benefits of ERP through organizational information processing theory. *Information & Management, 41*(4), 431–443.

Goerzen, A., & Beamish, P. W. (2003). Geographic scope and multinational enterprise performance. *Strategic Management Journal, 24*(13), 1289–1306.

Goerzen, A., & Beamish, P. W. (2005). The effect of alliance network diversity on multinational enterprise performance. *Strategic Management Journal, 26*(4), 333–354.

Golini, R., Deflorin, P., & Scherrer, M. (2016). Exploiting the potential of manufacturing network embeddedness. *International Journal of Operations & Production Management, 36*(12), 1741–1768.

Golini, R., & Gualandris, J. (2018). An empirical examination of the relationship between globalization, integration and sustainable innovation within manufacturing networks. *International Journal of Operations & Production Management, 38*(3), 874–894.

Green, W. (2017). *Third of UK firms plan to sever European supply chains—Supply management.* Available at: https://www.cips.org/supply-management/news/2017/may/third-of-uk-firms-plan-to-sever-european-supply-chains/. Accessed 18 Sept 2021.

Gu, M., Yang, L., & Huo, B. (2021). The impact of information technology usage on supply chain resilience and performance: An ambidexterous view. *International Journal of Production Economics, 232*, 107956.

Gupta, S., Kumar, S., Kamboj, S., Bhushan, B., & Luo, Z. (2019). Impact of IS agility and HR systems on job satisfaction: An organizational information processing theory perspective. *Journal of Knowledge Management, 23*(9), 1782–1805.

Gysegom, W., Sabah, R., Schlichter, M., Schmitz, C., & Soubien, F. (2019). *Brexit: The bigger picture—Rethinking supply chain strategies after Brexit.* McKinsey & Company. Available at: https://www.mckinsey.com/featured-insights/europe/brexit-the-bigger-picture-rethinking-supply-chains-in-a-time-of-uncertainty. Accessed 17 Sept 2021.

Habermann, M., Blackhurst, J., & Metcalf, A. Y. (2015). Keep your friends close? Supply chain design and disruption risk. *Decision Sciences, 46*(3), 491–526.

Hair, J. F., Black, W. C., Babin, B. J., & Anderson, R. E. (2010). *Multivariate data analysis* (7th edn.). Pearson.

Intel. (2020a). *How many manufacturing fabs does Intel have?* Available at https://www.intel.com/content/www/us/en/support/articles/000015142/programs.html. Accessed 21 Sept 2021.

Intel. (2020b). *Intel 2020–21 corporate responsibility report.* Available at http://csrreportbuilder.intel.com/pdfbuilder/pdfs/CSR-2020-21-Supply-Chain-Summary.pdf. Accessed 21 Sept 2021.

Jüttner, U., Peck, H., & Christopher, M. (2003). Supply chain risk management: Outlining an agenda for future research. *International Journal of Logistics: Research and Applications, 6*(4), 197–210.

Kauppi, K., Longoni, A., Caniato, F., & Kuula, M. (2016). Managing country disruption risks and improving operational performance: Risk management along integrated supply chains. *International Journal of Production Economics, 182*, 484–495.

Kim, K. K., Umanath, N. S., & Kim, B. H. (2005). An assessment of electronic information transfer in B2B supply-channel relationships. *Journal of Management Information Systems, 22*(3), 294–320.

Kim, Y. H., & Schoenherr, T. (2018). The effects of supply chain integration on the cost efficiency of contract manufacturing. *Journal of Supply Chain Management, 54*(3), 42–64.

Klassen, R. D., & Vachon, S. (2003). Collaboration and evaluation in the supply chain: The impact on plant-level environmental investment. *Production and Operations Management, 12*(3), 336–352.

Kleindorfer, P. R., & Saad, G. H. (2009). Managing disruption risks in supply chains. *Production and Operations Management, 14*(1), 53–68.

Lam, H. K. S., Yeung, A. C. L., & Cheng, T. C. E. (2016). The impact of firms' social media initiatives on operational efficiency and innovativeness. *Journal of Operations Management, 47–48*(1), 28–43.

Leuschner, R., Rogers, D. S., & Charvet, F. F. (2013). A meta-analysis of supply chain integration and firm performance. *Journal of Supply Chain Management, 49*(2), 34–57.

Li, G., Fan, H., Lee, P. K. C., & Cheng, T. C. E. (2015). Joint supply chain risk management: An agency and collaboration perspective. *International Journal of Production Economics, 164*, 83–94.

Li, Y., Dai, J., & Cui, L. (2020). The impact of digital technologies on economic and environmental performance in the context of industry 4.0: A moderated mediation model. *International Journal of Production Economics, 229*, 107777.

References

Liu, H., Wei, S., Ke, W., Wei, K. K., & Hua, Z. (2016). The configuration between supply chain integration and information technology competency: A resource orchestration perspective. *Journal of Operations Management, 44*(1), 13–29.

Liu, Y., Blome, C., Sanderson, J., & Paulraj, A. (2018). Supply chain integration capabilities, green design strategy and performance: A comparative study in the auto industry. *Supply Chain Management: an International Journal, 23*(5), 431–443.

Lu, G., & Shang, G. (2017). Impact of supply base structural complexity on financial performance: Roles of visible and not-so-visible characteristics. *Journal of Operations Management, 53–56*(1), 23–44.

Luo, Y. (2007). From foreign investors to strategic insiders: Shifting parameters, prescriptions and paradigms for MNCs in China. *Journal of World Business, 42*(1), 14–34.

Madhavan, R., Gnyawali, D. R., & He, J. (2004). Two's Company, three's a crowd? Triads in cooperative-competitive networks. *Academy of Management Journal, 47*(6), 918–927.

Mani, D., Barua, A., & Whinston, A. (2010). An empirical analysis of the impact of information capabilities design on business process outsourcing performance. *MIS Quarterly, 34*(1), 39–62.

Manuj, I., Esper, T. L., & Stank, T. P. (2014). Supply chain risk management approaches under different conditions of risk. *Journal of Business Logistics, 35*(3), 241–258.

Manuj, I., & Mentzer, J. T. (2008a). Global supply chain risk management. *Journal of Business Logistics, 29*(1), 133–155.

Manuj, I., & Mentzer, J. T. (2008b). Global supply chain risk management strategies. *International Journal of Physical Distribution & Logistics Management, 38*(3), 192–223.

Munir, M., Jajja, M. S. S., Chatha, K. A., & Farooq, S. (2020). Supply chain risk management and operational performance: The enabling role of supply chain integration. *International Journal of Production Economics, 227*, 107667.

Narasimhan, R., & Talluri, S. (2009). Perspectives on risk management in supply chains. *Journal of Operations Management, 27*(2), 114–118.

Nunnally, J. C. (1978). *Psychometric theory* (2nd ed.). McGraw-Hill.

Paulraj, A., Lado, A. A., & Chen, I. J. (2008). Inter-organizational communication as a relational competency: Antecedents and performance outcomes in collaborative buyer-supplier relationships. *Journal of Operations Management, 26*(1), 45–64.

Perrow, C. (1967). A framework for the comparative analysis of organizations. *American Sociological Review, 32*(2), 194.

Porter, M. E. (1985). *Competitive advantage: Creating and sustaining superior performance* (Vol. 43). Free Press.

Prajogo, D., & Olhager, J. (2012). Supply chain integration and performance: The effects of long-term relationships, information technology and sharing, and logistics integration. *International Journal of Production Economics, 135*(1), 514–522.

Premkumar, G., Ramamurthy, K., & Saunders, C. S. (2005). Information processing view of organizations: An exploratory examination of fit in the context of interorganizational relationships. *Journal of Management Information Systems, 22*(1), 257–294.

Safizadeh, M. H., Ritzman, L. P., & Mallick, D. (2000). Revisiting alternative theoretical paradigms in manufacturing strategy. *Production and Operations Management, 9*(2), 111–126.

Schoenherr, T., & Swink, M. (2012). Revisiting the arcs of integration: Cross-validations and extensions. *Journal of Operations Management, 30*(1–2), 99–115.

Shou, Y., Hu, W., Kang, M., Li, Y., & Park, Y. W. (2018a). Risk management and firm performance: The moderating role of supplier integration. *Industrial Management and Data Systems, 118*(7), 1327–1344.

Shou, Y., Kang, M., & Park, Y. (2022). *Supply chain integration for sustainable advantages.* Springer.

Shou, Y., Li, Y., Park, Y., & Kang, M. (2018b). Supply chain integration and operational performance: The contingency effects of production systems. *Journal of Purchasing and Supply Management, 24*(4), 352–360.

Sodhi, M. S., Son, B. G., & Tang, C. S. (2012). Researchers' perspectives on supply chain risk management. *Production and Operations Management, 21*(1), 1–13.

Srinivasan, R., & Swink, M. (2015). Leveraging supply chain integration through planning comprehensiveness: An organizational information processing theory perspective. *Decision Sciences, 46*(5), 823–861.

Srinivasan, R., & Swink, M. (2018). An investigation of visibility and flexibility as complements to supply chain analytics: An organizational information processing theory perspective. *Production and Operations Management, 27*(10), 1849–1867.

Stock, G. N., Greis, N. P., & Kasarda, J. D. (2000). Enterprise logistics and supply chain structure: The role of fit. *Journal of Operations Management, 18*(5), 531–547.

Stock, G. N., & Tatikonda, M. V. (2008). The joint influence of technology uncertainty and interorganizational interaction on external technology integration success. *Journal of Operations Management, 26*(1), 65–80.

Stock, G. N., Tsai, J. C. A., Jiang, J. J., & Klein, G. (2021). Coping with uncertainty: Knowledge sharing in new product development projects. *International Journal of Project Management, 39*(1), 59–70.

Su, Z., Xie, E., Liu, H., & Sun, W. (2013). Profiting from product innovation: The impact of legal, marketing, and technological capabilities in different environmental conditions. *Marketing Letters, 24*(3), 261–276.

Swink, M., Narasimhan, R., & Wang, C. (2007). Managing beyond the factory walls: Effects of four types of strategic integration on manufacturing plant performance. *Journal of Operations Management, 25*(1), 148–164.

Szász, L., Scherrer, M., & Deflorin, P. (2016). Benefits of internal manufacturing network integration. *International Journal of Operations & Production Management, 36*(7), 757–780.

Tang, C. S. (2006). Perspectives in supply chain risk management. *International Journal of Production Economics, 103*(2), 451–488.

Tang, O., & Nurmaya Musa, S. (2011). Identifying risk issues and research advancements in supply chain risk management. *International Journal of Production Economics, 133*(1), 25–34.

Tatikonda, M. V., & Montoya-Weiss, M. M. (2001). Integrating operations and marketing perspectives of product innovation: The influence of organizational process factors and capabilities on development performance. *Management Science, 47*(1), 151–172.

Thomé, A. M. T., & Sousa, R. (2016). Design-manufacturing integration and manufacturing complexity. *International Journal of Operations & Production Management, 36*(10), 1090–1114.

Tushman, M. L., & Nadler, D. A. (1978). Information processing as an integrating concept in organizational design. *Academy of Management Review, 3*(3), 613–624.

Ven, A. H. V., De, D., & A.L. and Koenig, R. (1976). Determinants of coordination modes within organizations. *American Sociological Review, 41*(2), 322.

Wang, E. T. G., Tai, J. C. F., & Grover, V. (2013). Examining the relational benefits of improved interfirm information processing capability in buyer-supplier dyads. *MIS Quarterly, 37*(1), 149–173.

Wieland, A., & Wallenburg, C. M. (2013). The influence of relational competencies on supply chain resilience: A relational view. *International Journal of Physical Distribution & Logistics Management, 43*(4), 300–320.

Wiengarten, F., Humphreys, P., Gimenez, C., & McIvor, R. (2016). Risk, risk management practices, and the success of supply chain integration. *International Journal of Production Economics, 171*, 361–370.

Wiengarten, F., & Longoni, A. (2018). How does uncertainty affect workplace accidents? Exploring the role of information sharing in manufacturing networks. *International Journal of Operations & Production Management, 38*(1), 295–310.

Williams, B. D., Roh, J., Tokar, T., & Swink, M. (2013). Leveraging supply chain visibility for responsiveness: The moderating role of internal integration. *Journal of Operations Management, 31*(7–8), 543–554.

References

Wong, C. W. Y., Lai, K., & Bernroider, E. W. N. (2015a). The performance of contingencies of supply chain information integration: The roles of product and market complexity. *International Journal of Production Economics, 165*, 1–11.

Wong, C. W. Y., Lai, K. H., Cheng, T. C. E., & Lun, Y. H. V. (2015b). The role of IT-enabled collaborative decision making in inter-organizational information integration to improve customer service performance. *International Journal of Production Economics, 159*, 56–65.

Wong, C. W. Y., Lirn, T. C., Yang, C. C., & Shang, K. C. (2020). Supply chain and external conditions under which supply chain resilience pays: An organizational information processing theorization. *International Journal of Production Economics, 226*, 107610.

Wong, C. Y., Boon-Itt, S., & Wong, C. W. Y. (2011). The contingency effects of environmental uncertainty on the relationship between supply chain integration and operational performance. *Journal of Operations Management, 29*(6), 604–615.

Yang, J., Xie, H., Yu, G., & Liu, M. (2021). Antecedents and consequences of supply chain risk management capabilities: An investigation in the post-coronavirus crisis. *International Journal of Production Research, 59*(5), 1573–1585.

Yu, W., Zhao, G., Liu, Q., & Song, Y. (2021).Role of big data analytics capability in developing integrated hospital supply chains and operational flexibility: An organizational information processing theory perspective. *Technological Forecasting and Social Change, 163*, 120417.

Zaheer, A., & Hernandez, E. (2011). The geographic scope of the MNC and its alliance portfolio: Resolving the paradox of distance. *Global Strategy Journal, 1*(1–2), 109–126.

Zeng, B., & Yen, B.P.-C. (2017). Rethinking the role of partnerships in global supply chains: A risk-based perspective. *International Journal of Production Economics, 185*, 52–62.

Zhou, C., & Li, J. (2008). Product innovation in emerging market-based international joint ventures: An organizational ecology perspective. *Journal of International Business Studies, 39*(7), 1114–1132.

Zhou, H., & Benton, W. C. (2007). Supply chain practice and information sharing. *Journal of Operations Management, 25*(6), 1348–1365.

Zhu, S., Song, J., Hazen, B. T., Lee, K., & Cegielski, C. (2018). How supply chain analytics enables operational supply chain transparency. *International Journal of Physical Distribution & Logistics Management, 48*(1), 47–68.

Chapter 7
Supply Chain Integration and Sustainability: The Supply Chain Learning Perspective

Abstract Sustainability management practices (SMPs) have attracted increasing attentions from supply chain researchers in recent years, whereas research on how firms collaborate with supply chain partners to implement SMPs successfully is still lacking. Drawing on the supply chain learning (SCL) perspective, we propose that supply chain integration (SCI) contributes to the successful implementation of SMPs. A structural equation modeling analysis is employed to test the proposed hypotheses using data collected from the International Manufacturing Strategy Survey (IMSS) project database. The findings suggest that supplier and customer integration are vital enablers for both intra- and inter-organizational SMPs. Moreover, both intra- and inter-organizational SMPs are verified to be significantly and positively associated with sustainability performance (i.e., economic, environmental and social performance) and function as complements to jointly enhance environmental and social performance simultaneously. This study employs the SCL perspective to incorporate SCI into the sustainability literature, providing a new perspective on sustainability and supply chain management research.

Keywords Supply chain integration · Supply chain learning · Sustainability management practices · Sustainability performance

7.1 Introduction

Over the past few decades, as firms have vigorously pursued competitive advantages within the turbulent global business environment, the importance of sustainability to a firm's bottom line has steadily grown (Kleindorfer et al., 2005; Lubin & Esty, 2010). Commensurate with the increasing importance of sustainability, supply chain researchers have examined whether extending sustainability issues into a firm's supply chain (which is beyond its internal operations) is a crucial step in improving its sustainability performance (Beske & Seuring, 2014; Seuring & Müller, 2008; Winter & Knemeyer, 2013). In this regard, given the complex and global nature

---

This chapter is a revised version of the following journal paper: Kang, M., Yang, M. G. M., Park, Y., & Huo, B. (2018). Supply chain integration and its impact on sustainability. *Industrial Management & Data Systems*, 118(9), 1749–1765.

© The Author(s), under exclusive license to Springer Nature Singapore Pte Ltd. 2022
Y. Shou et al., *Supply Chain Integration for Sustainable Advantages*,
https://doi.org/10.1007/978-981-16-9332-8_7

of sustainability issues, one of the expanding areas of interest has been whether companies understand the importance of collaboration within their supply chains (Seuring & Müller, 2008).

Despite the growing research on sustainable supply chain management (SSCM), it is still unclear how manufacturing firms cooperate with supply chain partners to achieve their desired sustainability performance. Several sustainability studies have examined external pressure as a main driver of firm sustainability (Dimaggio & Powell, 2000; Liu et al., 2010; Sarkis et al., 2010). However, little research has been conducted on how manufacturers adopt supply chain integration (SCI) activities to facilitate their sustainability management practices (SMPs) and enhance their sustainability performance. Research has investigated the positive influence of customer integration (Gelhard & von Delft, 2016), supply management capabilities (Bowen et al., 2001), and strategic purchasing in supply management (Paulraj, 2011) on SMPs or sustainability performance, indirectly representing a possible link between SCI and sustainability.

The supply chain learning (SCL) perspective emphasizes the role of learning through a firm's collaborative supply chain relationships in acquiring and sustaining competitive advantages. Several studies have employed this perspective to explain benefits of SCI. For example, Spekman et al. (2002) identified that key characteristics of SCI like integrative mechanisms, shared culture, commitment, trust and communications are pre-conditions for SCL and contribute to a firm's cost and revenue performance indirectly. Khan and Wisner (2019) suggested that the positive impact of SCI on supply chain agility is mediated by internal and external learning. However, sustainability knowledge obtained from the integration of the supply chain has been largely ignored in prior studies and whether a firm can benefit from learning through SCI in the implementation of SMPs remains uncovered.

The positive relationship between SCI and SMPs is also evident in the practical context. For instance, Nestlé, one of the world's largest food and beverage companies, started to develop modernizing dairy farms in China in response to the Chinese government's call for economic sustainability in the dairy industry since 2008. Nestlé emphasized supplier integration and developed Fresh Milk Procurement and Agriculture Service Departments specialized in communicating with suppliers (i.e., daily farmers in this example). Technical assistant (TA) supervisors in the department visited a certain number of farmers frequently and provided them training and technical assistance. Nestlé collected information about farmers' willingness to upgrade and their needs through their collaborative relationships with suppliers in the process of modernizing daily firms. Furthermore, TA supervisors informed farmers the latest government policy and educated them the urgency to upgrade. Thus, by implementing supplier integration, Nestlé could speed up its modernization process and achieve sustainability (Gong et al., 2018). Nestlé also faced internal challenges in its modernization process. Employees were resistant to the new production system since they had been working on the traditional model for more than 20 years. To educate employees new capabilities and skills, Nestlé relied on its internal integration capability to a large extent. Different internal departments shared information about modernizing dairy farms and developed a common understanding of the firm's sustainability development goal. In sum, we can infer from the Nestlé case that SCI is related positively and closely with the success of SMPs.

This study recognizes the need to further investigate the interrelationship between internal and external SCI and a firm's supply chain activities to improve sustainability. We seek to answer the following research questions: First, how do the three dimensions of SCI influence SMPs? Second, how do SMPs influence manufacturing firms' sustainability performance? By answering these important questions from a SCL perspective, this study, which extends sustainability research into supply chains, makes the following contributions. First, it provides new perspectives from which to understand the important role of SCI in improving intra- and inter-organizational SMPs and sustainability performance. Second, it provides practical insights into the ways manufacturers successfully implement SCI and SMPs to achieve their desired sustainability performance.

This chapter is structured as follows. In Sect. 2, we first review the literature about SCL, which is the theoretical basis of this study, and then, introduce what has been discussed about SMPs in prior studies. In this way, we identify research gaps and develop our hypotheses about the relationship between SCI, SMPs and sustainable performance. Section 3 elaborates on the data and variables of this study. Section 4 reports the results of data analyses. Explanations of our empirical results and contributions to the literature of this study are provided in Sect. 5. Finally, we discuss limitations and future research directions in Sect. 6.

7.2 Literature Review and Hypotheses Development

7.2.1 Supply Chain Learning

Knowledge is a crucial resource for a firm to acquire and sustain its competitive advantages since it is socially complex and difficult to imitate (Grant, 1996; Spender, 1996). Hence, it is increasingly important to understand organizational learning (OL), which refers to "the change in an organization's knowledge that occurs as a function of experience" (Argote & Miron-Spektor, 2011). There are two research streams in the existing OL literature: most of the early studies were concerned about intra-organizational learning while inter-organizational learning has received increasing attention in recent years (Coghlan & Coughlan, 2015; Yang et al., 2019). Intra-organizational learning focuses on identification of information needs, information sharing between different departments and organizational memory formulation within organizations (Day, 1994). Inter-organizational learning refers to the cross-boundary learning process between different individual organizations (Theodorakopoulos et al., 2005).

With the arisen prominence of an organization's supply chain management (SCM) since 1990s, learning in the supply chain context has increasingly become a critical concern for both academics and practitioners (Willis et al., 2016; Yang et al., 2019).

SCL derives and is extended from both intra-organizational and inter-organizational learning (Huo et al., 2020). It refers to the process of acquiring, assimilating and exploiting knowledge from all fragments of the SC composition to maximize value for the end customer (Huo et al., 2020). According to Kogut and Zander (1992), knowledge in SCL can be divided into two categories: explicit knowledge and tacit knowledge. Explicit knowledge can be transmitted without loss of integrity once rules for deciphering it are known. It usually includes facts, axiomatic propositions and symbols. By contrast, tacit knowledge is embedded in an organization's daily routines and usually difficult to codify. SCL is actually a dynamic process in which explicit knowledge and tacit knowledge are exchanged and transformed across different supply chain parties (Wu, 2008). There are four different supply chain knowledge creation modes: socialization mode in which tacit knowledge is shared through social interactions, externalization mode in which tacit knowledge is converted into explicit knowledge in formal, systematic language, combination mode in which explicit knowledge is transmitted through written media (e.g., reports and 3D CAD), and internalization mode in which explicit knowledge is transferred into tacit knowledge through practice (Wu, 2008).

The integration of the supply chain not only focuses on flows of tangible assets, but also emphasizes information flows (Prajogo & Olhager, 2012). Information sharing in SCI facilitates a firm's combination knowledge conversion process. Furthermore, SCI brings frequent and in-depth cross-boundary communication, which is the core mechanism for socialization knowledge creation. Besides, efficiency in knowledge internalization and externalization processes can be promoted by internal integration since the integration of internal functions enables an enterprise to combine information from different departments and develop contextualized beliefs effectively. In sum, SCI constitutes structural basis for knowledge creation along the supply chain.

Several prior studies have proved that benefits of SCI accrue from learning (Bessant et al., 2003; Flint et al., 2008; Manuj et al., 2013). Zhu et al. (2018) argued that the positive relationship between SCI and focal firm performance is mediated by SCL since the sharing of technological, market, manufacturing and inventory information across key suppliers and key customers is facilitated by SCI. Yu et al. (2013) also proved that SCL supported by integration enables the formulation of a common understanding of customer needs among supply chain parties. Thus, enterprises can ensure consistency between customer expectations and their supply chain capabilities.

Although the role of SCL in promoting an enterprise's competitive advantages has been identified in prior studies (Bessant et al., 2003), there is very little research focusing on sustainability knowledge learnt from supply chain parties (Gong et al., 2018). Implementing SMPs is a significant challenge for businesses. According to Smit et al. (2008), successful SMP implementation requires a firm to break its long-held cultural foundations and acquire information and know-how through stakeholder engagement and sharing of experience. Thus, learning is critical for an enterprise's pursuit of sustainability. A handful of studies have investigated how firms develop SSCM through generating learning processes. For instance, Oelze et al. (2016) employed a multiple in-depth case analysis to identify potential channels

7.2 Literature Review and Hypotheses Development

through which enterprises can learn how to implement sustainable policies in supply chains. Gong et al. (2018) investigated the SCL of sustainability process in multi-tier supply chains with the support of a focal firm's internal and external resources. The study conducted by Graham (2018) demonstrated that internal integration and learning capabilities play a critical role in extending environmental efforts to the supply chain level. However, little empirical research has been conducted on the relationship between learning through collaborations along the supply chain and sustainability management. Therefore, this study extends the existing literature and follows the guideline of SCL perspective to explain the relationship between SCI and the implementation of SMPs.

7.2.2 Sustainability Management Practices

Firms who want to improve sustainability pursue both intra- and inter-organizational SMPs (Wang & Dai, 2018).

Intra-organizational SMPs. Firms have increasingly adopted internal operational practices that enhance sustainability at the company level. One of these practices is environmental management, which aims to save costs related to environmental pollution (Lucas & Noordewier, 2016). Activities like reusing, recycling, and remanufacturing can minimize the negative environmental impact of waste disposal, extraction of raw materials, transportation, and distribution. Another internal operational practice enhances the social dimension of sustainability by improving employee well-being (Pagell & Gobeli, 2009; Voorde et al., 2012). Pagell and Gobeli (2009) reported that employee well-being practices were positively related to operational performance. Because the number of workplace accidents and injuries has steadily grown over the past decade, maintaining workplace safety is increasingly important for firms to protect employees and promote firm welfare (Das et al., 2008; Okun et al., 2016).

Inter-organizational SMPs. Companies' efforts to address sustainability have extended beyond their internal operations to helping suppliers meet sustainability standards that satisfy their customers' sustainability expectations. Incidents such as the Mattel toy recall and Unilever's palm oil contract suspension suggest that a supplier's failure to meet environmental standards can have a substantial negative impact on the focal company, causing immediate financial loss and long-term damage to the company's reputation (Zhang et al., 2011). Firms like Nike and Adidas have struggled to address suppliers' social equity issues such as inhumane working conditions (Seuring & Müller, 2008). These environmental and social problems have primarily come from supply chain partners who have been beyond their direct control. Thus, focal firms have progressively recognized the strategic importance of incorporating sustainability considerations in managing their major suppliers' performance (Gimenez & Tachizawa, 2012).

7.2.3 SCI and SMPs

Gimenez and Tachizawa (2012) suggested that both internal and external factors enabled firms to facilitate sustainability practices. They defined enablers as the "factors that assist firms in achieving sustainability practices" (p. 537). Internal enablers include but are not limited to the firm's environmental commitment, top management support, resource availability, purchasing personnel's supply management capability, and proper performance measurement systems. External enablers include supply chain related capabilities such as trust, national culture, and logistical and technological integration. Thus, both internal and external SCI could contribute to the implementation of SMPs since it requires collaborations of all parties along the supply chain (Pullman et al., 2009).

Supplier integration has been one of the key functional practices within the supply chain (Perols et al., 2013; Zhang et al., 2015). Managing suppliers through sharing of key information, involving suppliers in the product development process, and developing supplier programs has long been identified as a source for focal firms' competitive advantages (Li et al., 2005). Learning through collaborative activities with suppliers along the supply chain can help focal firms to identify multiple sustainability challenges (Huq et al., 2016; Klassen & Vachon, 2003). Drawing on the SCL perspective, we propose that suppliers can acquire information about the focal firms' sustainability development plans through the integration of the supply chain and modify their production technologies and procedures accordingly. A shared understanding of a firm's strategic sustainability goal across the supply chain is the basis for the smooth implementation of SMPs (Gong et al., 2018). Furthermore, supplier integration facilitates communications which lie at the heart of know-how transfer (Zhu et al., 2018). Hence, suppliers can learn tacit knowledge about how to achieve sustainability in their daily operations, contributing to the focal firm's implementation of SMPs as well.

Firms wanting to improve sustainability pursue both intra- and inter-organizational SMPs that enhance it. Because supplier integration is characterized by strategic collaboration, it may serve as a vital enabler in facilitating firms' successful implementation of intra- and inter-organizational SMPs. Thus, we suggest the following hypotheses.

H1a Supplier integration is positively associated with a firm's intra-organizational SMPs.

H1b Supplier integration is positively associated with a firm's inter-organizational SMPs.

A major driver of SMPs is the pressure from external stakeholders and customers (Paulraj, 2011). SCL through customer integration fosters capability development to serve the end customers (Van Wijk et al., 2008). Thus, firms that integrate customers into their operational and supply chain activities are more likely to learn customers' potential environmental and social concerns and implement SMPs accordingly to

7.2 Literature Review and Hypotheses Development 135

promote customer satisfaction (Gelhard & von Delft, 2016). Furthermore, a strong relationship between the focal firm and customers offers opportunities for co-recognizing innovation/market attributes that contribute to efficiencies in the implementation process of SMPs. Therefore, we infer that customer integration may play an important role in implementing both intra- and inter-organizational SMPs. The following hypotheses are thus posited.

H2a Customer integration is positively associated with a firm's intra-organizational SMPs.

H2b Customer integration is positively associated with a firm's Inter-organizational SMPs.

Internal integration focuses on breaking down functional barriers through activities, including information sharing, joint decision making, and cross-functional teamwork (Flynn et al., 2010). By improving horizontal linkages across internal functional units, internal integration fosters information sharing and collaboration between different functions (Antonio et al., 2009; Zhao et al., 2011). As companies implement SMPs, they will go through a process of learning since new technologies and capabilities are required (Hart & Dowell, 2011). Through cross-functional teamwork and collaboration, internal integration may play a central role in such learning process (Wolf, 2013, 2014). In addition, internal integration promotes the alignment of functional practices and goals with strategic business priorities (e.g., sustainability) (Narasimhan & Das, 2001). Such alignment may help to transform a firm's sustainability priorities into operational practices through cross-functional collaboration, rendering the organizational structure more suitable for sustainability. Therefore, we make the following hypotheses.

H3a Internal integration is positively associated with a firm's intra-organizational SMPs.

H3b Internal integration is positively associated with a firm's inter-organizational SMPs.

7.2.4 SMPs and Sustainability Performance

The positive link between internal environmental management practices and pertinent sustainability performance indicators has been well recognized (Pullman et al., 2009; Zhu & Sarkis, 2004). For example, environmental practices that allow an organization to design eco-friendly products can reduce environmental pollution and improve their sustainability performance (Hammouri et al., 2009). Similarly, environmental recycling practices help firms reuse, recycle, and remanufacture materials, components, and returned products, facilitating their environmental friendliness (Sarkis et al., 2010).

Employee health and safety systems aim to boost employee well-being (Pagell & Gobeli, 2009). The implementation of well-being practices is positively related to better overall sustainability performance outcomes (Rothenberg et al., 2001). Scholars have supported the notion that employee-related practices directly associated with positive employee attitudes lead to overall improvements in product quality, and thus, improve economic performance (Flynn et al., 1995). A firm's activities aimed at addressing safety issues in daily operations also can improve its social reputation (Pagell & Gobeli, 2009). Therefore, we contend that internal employee practices are positively associated with sustainability performance. We posit the following hypothesis.

H4 A firm's intra-organizational SMPs are positively associated with sustainability performance.

Using proper evaluation schemes to monitor and evaluate suppliers on whether they meet sustainability standards protects companies from potential risks related to environmental damage and violations of social standards (Koplin et al., 2007). Thus, monitoring may prevent unnecessary financial loss due to the high probability that supplier evaluations improve environmental performance and bring about positive economic performance. Furthermore, overseeing the qualifications of suppliers during the evaluation process helps organizations manage their reputations and corporate legitimacy (Bai & Sarkis, 2010), thereby improving social performance.

Developing the long-term capacity of suppliers to meet increasingly complex sustainability standards has been a smart solution to the rising level of supplier-related accidents in many countries. Firms that offer education and training programs to transfer knowledge corresponding to sustainability criteria are more likely to build mutual trust with their suppliers and enhance their sustainable performance. Engaging in joint activities with suppliers along the supply chain can help a company identify the multiple challenges that arise from sustainability issues, including those of an environmental and social nature (Yang et al., 2010). Although implementing practices to educate suppliers sustainability standards may require a prolonged timeframe and high costs, such activities bring sustainable power to a company, allowing it to deal with unexpected disruptions that may destroy its entire supply chain. Therefore, we make the following hypothesis.

H5 A firm's inter-organizational SMPs are positively associated with sustainability performance.

7.3 Method

7.3.1 Data

To empirically test the proposed hypotheses, data collected from the six round International Manufacturing Strategy Survey (IMSS) were used. Details about the IMSS database are available in Chapter 1 of this book (Shou et al., 2022). After dropping responses with over 60% missing data, a final sample of 931 responses was extracted.

7.3.2 Measures

Measurements for the constructs in this study were adapted from the extant literature. Detailed survey questions are available in the Appendix of this book (Shou et al., 2022).

In this study, SCI includes customer, supplier, and internal integration. Based on previous SCI studies (Quesada et al., 2008; Sun & Ni, 2012; Wiengarten et al., 2014; Yang et al., 2016), customer and supplier integration were measured by collaborative approaches, information sharing, joint decision making, and system coupling with key customers and suppliers. Internal integration was measured by joint decision making and information sharing with purchasing and sales departments according to Yang et al. (2016).

This study classified a firm's intra-organizational SMPs and inter-organizational SMPs as two forms of SMP. Based on the work of Pagell and Gobeli (2009), intra-organizational SMP was operationalized including the environmental and employee well-being aspects of SMP. Inter-organizational SMPs were measured by supplier assessment and collaboration in sustainability issues, according to the work of Gualandris and Kalchschmidt (2014) and Wilding et al. (2012).

Following the triple bottom line (TBL) perspective proposed by Elkington (2010) and Carter and Rogers (2008), sustainability performance was operationalized as a second-order construct including environmental, social, and economic performance. Environmental performance was measured by indicators covering (1) materials, water, and/or energy consumption and (2) pollution emission and waste production levels (Golini et al., 2014; Paulraj, 2011). Social performance was measured by indicators covering (1) worker motivation and satisfaction and (2) health and safety conditions (Carter & Jennings, 2002; Golini et al., 2014; Paulraj, 2011). Finally, economic performance was measured across the dimensions of cost and time (Gimenez et al., 2012) with three items as one item (T2) was dropped.

Firm size was included as a control variable. In general, large firms with resource availability and business process capabilities tend to be better at implementing SMPs and achieving sustainability performance than small firms (Gualandris & Kalchschmidt, 2014). We measured firm size by the natural logarithm of the number of employees.

7.3.3 Reliability and Validity

The confirmatory factor analysis (CFA) verification of our measurement model was deemed successful with goodness-of-fit indices of acceptable values ($\chi^2 = 979.265$, df $= 247$, GFI $= 0.921$, CFI $= 0.940$ and RMSEA $= 0.05$) (Hu & Bentler, 1999; Schumacker & Lomax, 2004). The loadings for the first-order measurement model ranged from 0.607 to 0.881, providing evidence of convergent validity. With regard to the internal consistency of our constructs, the Cronbach's α and composite reliability (CR) values ranged from 0.684 to 0.890 and from 0.803 to 0.893, respectively, indicating an adequate level of internal consistency.

According to the test suggested by Zait and Bertea (2011), we assessed discriminant validity by testing whether the square root of the average variance extracted (AVE) value for each construct was greater than the correlation between the latent variables (Fornell & Larcker, 1981). Table 7.1 presents the means, standard deviations, and correlations between the constructs and the AVE. The results show that all of the square roots of the AVE were larger than the corresponding correlations, providing strong support for discriminant validity.

Finally, following the guidance of Marsh and Dennis (1985), we computed the target coefficient (T) that was a ratio of the chi-square value of the first-order model to the chi-square value of the second-order model. The T coefficient was 0.93, indicating that the second-order factor explained the majority of relationships between the first-order factors (i.e., environmental performance, social performance and economic performance). All of the second-order factor loadings were significant, providing further evidence that the second-order factor model was appropriate.

Table 7.1 Descriptive statistics, inter-construct correlations, and AVEs

	Mean	S.D	1	2	3	4	5	6	7	8
1. Supplier integration	3.04	0.860	**0.56**							
2. Customer integration	2.94	0.937	0.68[b]	**0.60**						
3. Internal integration	3.55	0.859	0.40[b]	0.39[b]	**0.68**					
4. Intra-organizational SMPs	3.34	0.942	0.39[b]	0.39[b]	0.43[b]	**0.58**				
5. Inter-organizational SMPs	2.80	1.041	0.46[b]	0.45[b]	0.37[b]	0.61[b]	**0.63**			
6. Environmental performance	3.19	0.591	0.14[b]	0.12[b]	0.15[b]	0.24[b]	0.27[b]	**0.71**		
7. Social performance	3.41	0.645	0.25[b]	0.24[b]	0.26[b]	0.38[b]	0.36[b]	0.32[b]	**0.68**	
8. Economic performance	3.08	0.563	0.14[b]	0.08[a]	0.15[b]	0.17[b]	0.21[b]	0.40[b]	0.20[b]	**0.62**
9. Firm size	6.03	1.720	0.20[b]	0.15[b]	0.11[b]	0.25[b]	0.18[b]	0.05	0.05	0.04

Note Values on the diagonal indicate AVE. [a]$p < 0.05$, [b]$p < 0.01$

7.4 Results

7.4.1 Structural Model Results

We examined the proposed relationship between the constructs by using structural equation modeling (SEM) with the maximum likelihood estimation method. The model fit indices of our structural model are shown as follows: $\chi^2 = 1269.518$, df = 282, GFI = 0.905, CFI = 0.919 and RMSEA = 0.061, indicating a good model fit. The results of the standardized path coefficients for each hypothesized causal relationship are provided in Fig. 7.1.

H1a and H1b predicted the positive effects of supplier integration on intra- and inter-organizational SMPs. In the SEM, the paths between both relationships were positive and significant (path coefficient = 0.447, p-value < 0.001 for H1a; path coefficient = 0.582, p-value < 0.001 for H1b), providing strong support for H1a and H1b. H2a and H2b, which predicted the positive effects of customer integration on intra- and inter-organizational SMPs, were also supported (path coefficient = 0.110, p-value < 0.05 for H2a; path coefficient = 0.147, p-value < 0.01 for H2b). H3a and H3b predicted the positive effects of internal integration on intra- and inter-organizational SMPs. The results show that the relationship between internal integration and intra-organizational SMPs was positive and significant (path coefficient = 0.150, p-value < 0.001), whereas no statistically significant relationship was found between internal integration and inter-organizational SMPs, supporting H3a but not H3b.

We examined differences in the strength of associations between SCI and SMPs. A relative effect analysis was conducted by comparing the chi-square difference between the constrained and unconstrained models. The results show that supplier integration had a significantly stronger effect on intra-organizational SMPs than

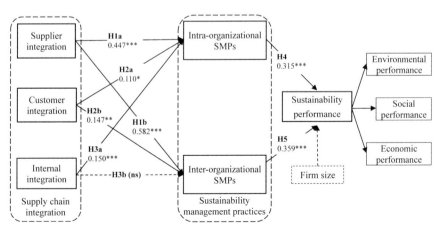

*p<0.05; **p<0.01; ***p<0.001; The dashed lines indicate non-significant paths.

Fig. 7.1 SEM results

both customer integration ($\Delta\chi = 20.72$, $\Delta df = 1$) and internal integration ($\Delta\chi = 44.18$, $\Delta df = 1$) and further confirm that the effect of supplier integration on the inter-organizational SMPs was significantly higher than customer integration ($\Delta\chi = 11.23$, $\Delta df = 1$).

H4 and H5 predicted that both intra- and inter-organizational SMPs would be positively associated with sustainability performance. In the structural equation model, the paths between both relationships were positive and significant (path coefficient = 0.315, p-value < 0.001 for H4; path coefficient = 0.359, p-value < 0.001 for H5), providing strong support for H4 and H5.

7.4.2 Additional Analyses

To further understand the relationships between SMPs and sustainability performance, we tested the individual effects of SMPs on each dimension of sustainability performance (environmental, social, and economic). Table 7.2 shows that there were significant direct relationships between both intra- and inter-organizational SMPs and environmental performance. Additionally, significant direction relationships were found between both intra- and inter-organizational SMPs and social performance. However, when it came to economic performance, only the coefficient for inter-organizational SMPs was statistically significant, indicating that intra-organizational SMPs were not directly related to economic performance. Using hierarchical regression analysis, we then tested the interaction effects of intra-organizational SMPs and inter-organizational SMPs on all three dimensions of sustainability performance. As shown in Table 7.2, the results reveal that the interaction terms of intra-organizational

Table 7.2 Hierarchical regression analyses

	Environmental performance			Social performance		Economic performance	
	Model 1		Model 2	Model 3	Model 4	Model 5	Model 6
Constant	2.646		2.605	2.570	2.482	2.701	2.684
Firm size	−0.007		−0.006	−0.020	−0.019	−0.002	−0.002
Intra-organizational SMPs	0.081[b]		0.092[b]	0.182[a]	0.206[a]	0.038	0.043[+]
Inter-organizational SMPs	0.112[a]		0.104[a]	0.127[a]	0.109[a]	0.093[a]	0.089[a]
Intra-organizational SMPs × Inter-organizational SMPs			0.038[c]		0.082[a]		0.016
R^2	0.084		0.089	0.170	0.188	0.046	0.047
Adjusted R^2	0.081		0.085	0.168	0.184	0.043	0.043
F	28.40[a]		22.51[a]	63.45[a]	53.43[a]	15.04[a]	11.48[a]

Notes [a]$p < 0.001$, [b]$p < 0.01$, [c]$p < 0.05$, [+]$p < 0.1$

7.4 Results   141

SMSs and inter-organizational SMPs were significantly correlated with environmental performance (path coefficient = 0.038, p-value < 0.05) and social performance (path coefficient = 0.082, p-value < 0.001), but not with economic performance (path coefficient = 0.016, p-value > 0.1). These results indicate that both intra-organizational SMPs and inter-organizational SMPs acted as complements to jointly enhance environmental and social performance.

7.5 Discussion

7.5.1 Findings

From the SCL perspective, this study identifies the role of SCI as an enabler that a firm may use to promote the successful implementation of SMPs. Our results reveal that both external integration (i.e., supplier and customer integration) and internal integration contribute to the implementation of intra-organizational SMPs. However, regarding inter-organizational SMPs, only the enabling effect of external integration is observed whereas that of internal integration is not. Our results are similar to Vachon and Klassen (2006), who emphasized that technological integration (defined as information and knowledge sharing with suppliers and customers taking place in strategic areas) was positively linked to environmental monitoring and collaboration activities with external supply chain partners. Taking the findings of the previous studies and our findings together, it is evident that successful inter-organizational SMPs require well-designed and implemented external SCI but do not rely on internal integration.

In addition to the foregoing, our findings indicate that supplier integration has a stronger positive impact on both intra- and inter-organizational SMPs than do customer and internal integration. This suggests that supplier integration with the three types of SCI is the most important enabler of SMPs. Undoubtedly, pressure from the customer side is a major driver of SMPs, as a number of studies have emphasized (Gualandris & Kalchschmidt, 2014; Sancha et al., 2015; Tate et al., 2010; Zhu et al., 2013). Customer requirements for sustainability are the key starting point for firms' SMPs efforts and drive firms to adopt sustainability-oriented strategies. However, they may be less helpful in putting sustainability-oriented strategies into daily practice than supplier integration. Contrary to the customers' driving role in SMPs, suppliers may be less likely to relate to a firm's sustainability-oriented strategies but nevertheless play a very important role in implementing SMPs. Studies have emphasized the importance of supply management and close cooperation with key suppliers when implementing SMPs (Bowen et al., 2001; Wilding et al., 2012), indicating the need for supplier integration. Similarly, our findings highlight the positive role of supplier integration in implementing SMPs, which we argue is greater than customer and internal integration.

Finally, the positive impact of both intra- and inter-organizational SMPs on sustainability performance (H4 and H5) were supported in our analysis. In other words, consistent with the findings of previous studies (Gimenez et al., 2012; Paulraj, 2011; Wilding et al., 2012), we found that both intra- and inter-organizational SMPs played a very important role in achieving the desired sustainability performance (i.e., economic, environmental and social performance). Furthermore, the interaction of intra- and inter-organizational SMPs was significantly associated with two sustainability performance measures: environmental and social performance. Such a positive interaction effect was not found for economic performance. In other words, both intra- and inter-organizational SMPs functioned as complements to jointly enhance environmental and social performance.

7.5.2 *Theoretical Implications*

This study makes contributions to the extant literature. First, it extends the sustainability management literature by revealing the role and effective use of SCI in intra- and inter-organizational SMPs to achieve desired sustainability performance. Most studies have emphasized external pressure (e.g., stakeholder and institutional) as the motivation underpinning sustainability management (Sancha et al., 2015; Tate et al., 2010; Wolf, 2014; Zhu et al., 2013). Only several studies have examined the impact of firm-specific resources and capabilities on SMPs and performance, and these have considered supply management capabilities (Bowen et al., 2001) and strategic purchasing (Paulraj, 2011). By incorporating SCI into sustainability management and investigating the three dimensions of SCI as enablers of SMPs, this study provides a new perspective on how firms implement SMPs to enhance sustainability performance.

Second, our study advances the SCL literature by providing empirical evidence that sustainability knowledge generation and transfer can be facilitated by the integration of the supply chain. In response to calls for integration of insights from organizational learning theory and supply chain management (Ketchen & Hult, 2007), several researchers have examined the mediating role of SCL in the relationship between SCI and operational performance (Khan & Wisner, 2019; Zhu et al., 2018). However, there is a lack of empirical studies investigating how learning through SCI promotes firm performance in a sustainable management context. Our results indicate that SCI enables focal firms to learn customers' environmental and social concerns effectively and develop sustainable skills and capabilities along the supply chain, which deserves more in-depth studies in the future.

7.6 Conclusions

With the growing importance of collaborating with supply chain partners to implement sustainability, this study contributes to the literature by providing SCL perspective on the important enablers of SMPs and valuable insights into the effective ways to use SCI in the sustainability management context. In this study, supplier and customer integration were enablers for both intra- and inter-organizational SMPs, whereas internal integration only enhanced intra-organizational SMPs. Intra- and inter-organizational SMPs were not only positively related to all three aspects of sustainability performance (i.e., economic, environmental and social performances), but also functioned complementarily to jointly promote both environmental and social performance. With that in mind, some limitations and issues must be further investigated.

First, although we explored the relationship between SCI and SMPs, future empirical investigations are needed to further examine this relationship. For example, the effects of SCI on SMPs may be influenced by various contingency factors, such as product complexity, supply chain complexity, and environmental uncertainties. Second, further investigation of the relationship between internal integration and inter-organizational SMPs is merited. Although our results show an insignificant relationship between internal integration and inter-organizational SMPs, we do not necessarily confirm that internal integration is useless to inter-organizational SMPs. Third, the inter-organizational SMPs in our study referred only to SMPs with suppliers. Studies have also considered sustainable supply management in the context of inter-organizational SMPs (Gualandris & Kalchschmidt, 2014; Paulraj, 2011; Wilding et al., 2012). However, it may be more ideal to include key customers in inter-organizational SMPs, thereby providing greater insight into understanding the enablers of SMPs.

Acknowledgements This work was supported by the National Natural Science Foundation of China under Grant Numbers 71525005 and 71372058.

References

Antonio, K. W. L., Richard, C. M. Y., & Tang, E. (2009). The complementarity of internal integration and product modularity: An empirical study of their interaction effect on competitive capabilities. *Journal of Engineering and Technology Management, 26*(4), 305–326.

Argote, L., & Miron-Spektor, E. (2011). Organizational learning: From experience to knowledge. *Organization Science, 22*(5), 1123–1137.

Bai, C., & Sarkis, J. (2010). Integrating sustainability into supplier selection with grey system and rough set methodologies. *International Journal of Production Economics, 124*(1), 252–264.

Beske, P., & Seuring, S. (2014). Putting sustainability into supply chain management. *Supply Chain Management, 19*(3), 322–331.

Bessant, J., Kaplinsky, R., & Lamming, R. (2003). Putting supply chain learning into practice. *International Journal of Operations and Production Management, 23*(2), 167–184.

Bowen, F. E., Cousins, P. D., Lamming, R. C., & Farukt, A. C. (2001). The role of supply management capabilities in green supply. *Production & Operations Management, 10*(2), 174–189.

Carter, C. R., & Jennings, M. M. (2002). Logistics social responsibility: An integrative framework. *Journal of Business Logistics, 23*(1), 145–180.

Carter, C. R., & Rogers, D. S. (2008). A framework of sustainable supply chain management: Moving toward new theory. *International Journal of Physical Distribution & Logistics Management, 38*(5), 360–387.

Coghlan, D., & Coughlan, P. (2015). Effecting change and learning in networks through network action learning. *Journal of Applied Behavioral Science, 51*(3), 375–400.

Das, A., Pagell, M., Behm, M., & Veltri, A. (2008). Toward a theory of the linkages between safety and quality. *Journal of Operations Management, 26*(4), 521–535.

Day, G. S. (1994). The capabilities of market-driven organizations. *Journal of Marketing, 58*(4), 37–52.

DiMaggio, P. J., & Powell, W. W. (2000). The iron cage revisited: Institutional isomorphism and collective rationality in organizational fields. *Advances in Strategic Management, 48*(2), 147–160.

Elkington, J. (2010). Cannibals with forks: The triple bottom line of 21st century business. *Environmental Quality Management, 8*(1), 37–51.

Flint, D. J., Larsson, E., & Gammelgaard, B. (2008). Exploring processes for customer value insights, supply chain learning and innovation: An international study. *Journal of Business Logistics, 29*(1), 257–281.

Flynn, B. B., Huo, B., & Zhao, X. (2010). The impact of supply chain integration on performance: A contingency and configuration approach. *Journal of Operations Management, 28*(1), 58–71.

Flynn, B. B., Schroeder, R. G., & Sakakibara, S. (1995). The impact of quality management practices on performance and competitive advantage. *Decision Sciences, 26*(5), 659–691.

Fornell, C., & Larcker, D. F. (1981). Structural equation models with unobservable variables and measurement error: Algebra and statistics. *Journal of Marketing Research, 18*(1), 382–388.

Gelhard, C., & von Delft, S. (2016). The role of organizational capabilities in achieving superior sustainability performance. *Journal of Business Research, 69*(10), 4632–4642.

Gimenez, C., Sierra, V., & Rodon, J. (2012). Sustainable operations: Their impact on the triple bottom line. *International Journal of Production Economics, 140*(1), 149–159.

Gimenez, C., & Tachizawa, E. M. (2012). Extending sustainability to suppliers: A systematic literature review. *Supply Chain Management, 17*(5), 531–543.

Golini, R., Longoni, A., & Cagliano, R. (2014). Developing sustainability in global manufacturing networks: The role of site competence on sustainability performance. *International Journal of Production Economics, 147*(Part B), 448–459.

Gong, Y., Jia, F., & Brown, S. (2018). Modernisation of dairy farms: The case of Nestlé's Dairy Farming Institute in China. *Emerald Emerging Markets Case Studies, 8*(1), 1–20.

Gong, Y., Jia, F., Brown, S., & Koh, L. (2018). Supply chain learning of sustainability in multi-tier supply chains: A resource orchestration perspective. *International Journal of Operations & Production Management, 38*(4), 1061–1090.

Graham, S. (2018). Antecedents to environmental supply chain strategies: The role of internal integration and environmental learning. *International Journal of Production Economics, 197*(3), 283–296.

Grant, R. M. (1996). Toward a knowledge-based theory of the firm. *Strategic Management Journal, 17,* 109–122.

Gualandris, J., & Kalchschmidt, M. (2014). Customer pressure and innovativeness: Their role in sustainable supply chain management. *Journal of Purchasing and Supply Management, 20*(2), 92–103.

Hammouri, H., Kabore, P., & Kinnaert, M. (2009). Adopting and applying eco-design techniques: A practitioners perspective. *Journal of Cleaner Production, 17*(5), 549–558.

Hart, S. L., & Dowell, G. (2011). A natural-resource-based view of the firm: Fifteen years after. *Journal of Management, 37*(5), 1464–1479.

References

Hu, L. T., & Bentler, P. M. (1999). Cutoff criteria for fit indexes in covariance structure analysis: Conventional criteria versus new alternatives. *Structural Equation Modeling: A Multidisciplinary Journal, 6*(1), 1–55.

Huo, B., Haq, M. Z. U., & Gu, M. (2020). The impact of IT application on supply chain learning and service performance. *Industrial Management and Data Systems, 120*(1), 1–20.

Huq, F. A., Chowdhury, I. N., & Klassen, R. D. (2016). Social management capabilities of multinational buying firms and their emerging market suppliers: An exploratory study of the clothing industry. *Journal of Operations Management, 46,* 19–37.

Ketchen, D. J., & Hult, G. T. M. (2007). Toward greater integration of insights from organization theory and supply chain management. *Journal of Operations Management, 25*(2), 455–458.

Khan, H., & Wisner, J. D. (2019). Supply chain integration, learning, and agility: Effects on performance. *Journal of Operations and Supply Chain Management, 12*(1), 14–23.

Klassen, R. D., & Vachon, S. (2003). Collaboration and evaluation in the supply chain: The impact on plant-level environmental investment. *Production & Operations Management, 12*(3), 336–352.

Kleindorfer, P. R., Singhal, K., & Van Wassenhove, L. N. (2005). Sustainable operations management. *Production and Operations Management, 14*(4), 482–492.

Kogut, B., & Zander, U. (1992). Knowledge of the firm, combinative capabilities, and the replication of technology. *Organization Science, 3*(3), 383–397.

Koplin, J., Seuring, S., & Mesterharm, M. (2007). Incorporating sustainability into supply management in the automotive industry—The case of the Volkswagen AG. *Journal of Cleaner Production, 15*(11–12), 1053–1062.

Large, R. O., Kramer, N., & Hartmann, R. K. (2013). Procurement of logistics services and sustainable development in Europe: Fields of activity and empirical results. *Journal of Purchasing and Supply Management, 19*(3), 122–133.

Li, S., Rao, S. S., Ragu-Nathan, T. S., & Ragu-Nathan, B. (2005). Development and validation of a measurement instrument for studying supply chain management practices. *Journal of Operations Management, 23*(6), 618–641.

Liu, H., Ke, W., Wei, K. K., Gu, J., & Chen, H. (2010). The role of institutional pressures and organizational culture in the firm's intention to adopt internet-enabled supply chain management systems. *Journal of Operations Management, 28*(5), 372–384.

Lubin, D. A., & Esty, D. C. (2010). The sustainability imperative. *Harvard Business Review, 88*(5), 42–50.

Lucas, M. T., & Noordewier, T. G. (2016). Environmental management practices and firm financial performance: The moderating effect of industry pollution-related factors. *International Journal of Production Economics, 175,* 24–34.

Manuj, I., Omar, A., & Yazdanparast, A. (2013). The quest for competitive advantage in global supply chains: The role of interorganizational learning. *Transportation Journal, 52*(4), 463–492.

Marsh, H. W., & Dennis, H. (1985). Application of confirmatory factor analysis to the study of self-concept: First- and higher order factor models and their invariance across groups. *Psychological Bulletin, 97*(3), 562–582.

Narasimhan, R., & Das, A. (2001). The impact of purchasing integration and practices on manufacturing performance. *Journal of Operations Management, 19*(5), 593–609.

Oelze, N., Hoejmose, S. U., Habisch, A., & Millington, A. (2016). Sustainable development in supply chain management: The role of organizational learning for policy implementation. *Business Strategy and the Environment, 25*(4), 241–260.

Okun, A. H., Guerin, R. J., & Schulte, P. A. (2016). Foundational workplace safety and health competencies for the emerging workforce. *Journal of Safety Research, 59,* 43–51.

Pagell, M., & Gobeli, D. (2009). How plant managers' experiences and attitudes toward sustainability relate to operational performance. *Production and Operations Management, 18*(3), 278–299.

Paulraj, A. (2011). Understanding the relationships between internal resources and capabilities, sustainable supply management and organizational sustainability. *Journal of Supply Chain Management, 47*(1), 19–37.

Perols, J., Zimmermann, C., & Kortmann, S. (2013). On the relationship between supplier integration and time-to-market. *Journal of Operations Management, 31*(3), 153–167.

Prajogo, D., & Olhager, J. (2012). Supply chain integration and performance: The effects of long-term relationships, information technology and sharing, and logistics integration. *International Journal of Production Economics, 135*(1), 514–522.

Pullman, M. E., Maloni, M. J., & Carter, C. R. (2009). Food for thought: Social versus environmental sustainability practices and performance outcomes. *Journal of Supply Chain Management, 45*(4), 38–54.

Quesada, G., Rachamadugu, R., Gonzalez, M., & Martinez, J. L. (2008). Linking order winning and external supply chain integration strategies. *Supply Chain Management: An International Journal, 13*(4), 296–303.

Rothenberg, S., Pil, F. K., & Maxwell, J. (2001). Lean, green, and the quest for superior environmental performance. *Production & Operations Management, 10*(3), 228–243.

Sancha, C., Longoni, A., & Giménez, C. (2015). Sustainable supplier development practices: Drivers and enablers in a global context. *Journal of Purchasing and Supply Management, 21*(2), 95–102.

Sarkis, J., Gonzalez-Torre, P., & Adenso-Diaz, B. (2010). Stakeholder pressure and the adoption of environmental practices: The mediating effect of training. *Journal of Operations Management, 28*(2), 163–176.

Schumacker, R. E., & Lomax, R. G. (2004). A beginner's guide to structural equation modeling. *Technometrics, 47*(4), 522–522.

Seuring, S., & Müller, M. (2008). From a literature review to a conceptual framework for sustainable supply chain management. *Journal of Cleaner Production, 16*(15), 1699–1710.

Shou, Y., Kang, M., & Park, Y. (2022). *Supply chain integration for sustainable advantages.* Springer.

Smit, A. A. H., Driessen, P. P. J., & Glasbergen, P. (2008). Constraints on the conversion to sustainable production: The case of the Dutch potato chain. *Business Strategy and the Environment, 17*(6), 369–381.

Spekman, R. E., Spear, J., & Kamauff, J. (2002). Supply chain competency: Learning as a key component. *Supply Chain Management, 7*(1), 41–55.

Spender, J. C. (1996). Making knowledge the basis of a dynamic theory of the firm. *Strategic Management Journal, 17*(S2), 45–62.

Sun, H., & Ni, W. (2012). The impact of upstream supply and downstream demand integration on quality management and quality performance. *International Journal of Quality & Reliability Management, 29*(8), 872–890.

Tate, W. L., Ellram, L. M., & Kirchoff, J. F. (2010). Corporate social responsibility reports: A thematic analysis related to supply chain management. *Journal of Supply Chain Management, 46*(1), 19–44.

Theodorakopoulos, N., Ram, M., Shah, M., & Boyal, H. (2005). Experimenting with supply chain learning (SCL): Supplier diversity and ethnic minority businesses. *International Entrepreneurship and Management Journal, 1*(4), 461–478.

Vachon, S., & Klassen, R. D. (2006). Extending green practices across the supply chain. *International Journal of Operations & Production Management, 26*(7), 795–821.

Van Wijk, R., Jansen, J. J. P., & Lyles, M. A. (2008). Inter- and intra-organizational knowledge transfer: A meta-analytic review and assessment of its antecedents and consequences. *Journal of Management Studies, 45*(4), 830–853.

Voorde, K. V. D., Paauwe, J., & Veldhoven, M. V. (2012). Employee well-being and the HRM–organizational performance relationship: A review of quantitative studies. *International Journal of Management Reviews, 14*(4), 391–407.

Wang, J., & Dai, J. (2018). Sustainable supply chain management practices and performance. *Industrial Management and Data Systems, 118*(1), 2–21.

Wiengarten, F., Pagell, M., Ahmed, M. U., & Gimenez, C. (2014). Do a country's logistical capabilities moderate the external integration performance relationship? *Journal of Operations Management, 32*(1–2), 51–63.

References

Wilding, R., Wagner, B., Gimenez, C., & Tachizawa, E. M. (2012). Extending sustainability to suppliers: A systematic literature review. *Supply Chain Management: An International Journal, 17*(5), 531–543.

Willis, G., Genchev, S. E., & Chen, H. (2016). Supply chain learning, integration, and flexibility performance: An empirical study in India. *International Journal of Logistics Management, 27*(3), 755–769.

Winter, M., & Knemeyer, A. M. (2013). Exploring the integration of sustainability and supply chain management: Current state and opportunities for future inquiry. *International Journal of Physical Distribution & Logistics Management, 43*(1), 18–38.

Wolf, J. (2013). Improving the sustainable development of firms: The role of employees. *Business Strategy and the Environment, 22*(2), 92–108.

Wolf, J. (2014). The relationship between sustainable supply chain management, stakeholder pressure and corporate sustainability performance. *Journal of Business Ethics, 119*(3), 317–328.

Wu, C. (2008). Knowledge creation in a supply chain. *Supply Chain Management, 13*(3), 241–250.

Yang, C., Chaudhuri, A., & Farooq, S. (2016). Interplant coordination, supply chain integration, and operational performance of a plant in a manufacturing network: A mediation analysis. *Supply Chain Management, 21*(5), 550–568.

Yang, C. L., Lin, S. P., Chan, Y. H., & Sheu, C. (2010). Mediated effect of environmental management on manufacturing competitiveness: An empirical study. *International Journal of Production Economics, 123*(1), 210–220.

Yang, M. G., Hong, P., & Modi, S. B. (2011). Impact of lean manufacturing and environmental management on business performance: An empirical study of manufacturing firms. *International Journal of Production Economics, 129*(2), 251–261.

Yang, Y., Jia, F., & Xu, Z. (2019). Towards an integrated conceptual model of supply chain learning: An extended resource-based view. *Supply Chain Management, 24*(2), 189–214.

Yu, W., Jacobs, M. A., Salisbury, W. D., & Enns, H. (2013). The effects of supply chain integration on customer satisfaction and financial performance: An organizational learning perspective. *International Journal of Production Economics, 146*(1), 346–358.

Zait, A., & Bertea, P. E. (2011). Methods for testing discriminant validity. *Management and Marketing Journal, 9*(2), 217–224.

Zhang, M., Zhao, X., Voss, C., & Zhu, G. (2015). Innovating through services, co-creation and supplier integration: Cases from China. *International Journal of Production Economics, 171*(2), 289–300.

Zhang, X., Shen, L., & Wu, Y. (2011). Green strategy for gaining competitive advantage in housing development: A China study. *Journal of Cleaner Production, 19*(2–3), 157–167.

Zhao, X., Huo, B., Selen, W., & Yeung, J. H. Y. (2011). The impact of internal integration and relationship commitment on external integration. *Journal of Operations Management, 29*(1), 17–32.

Zhu, Q., & Sarkis, J. (2004). Relationships between operational practices and performance among early adopters of green supply chain management practices in Chinese manufacturing enterprises. *Journal of Operations Management, 22*(3), 265–289.

Zhu, Q., Krikke, H., & Caniëls, M. C. J. (2018). Supply chain integration: Value creation through managing inter-organizational learning. *International Journal of Operations & Production Management, 38*(1), 211–229.

Zhu, Q., Sarkis, J., & Lai, K.-H. (2013). Institutional-based antecedents and performance outcomes of internal and external green supply chain management practices. *Journal of Purchasing and Supply Management, 19*(2), 106–117.

Chapter 8
Conclusions and Implications

Abstract In this chapter, we summarize this book by highlighting the findings and contributions of the research models proposed and tested in previous chapters. Managerial implications are elaborated in the hope of providing new insights to supply chain managers.

Keywords Supply chain integration · Finding · Implication

8.1 A Brief Summary

Supply chain integration (SCI) has been a hot topic in both academia and industry for decades. SCI is viewed as a highly effective and strategic supply chain management method, which explains the reason why SCI is so influential. In this book, the authors first give a systematic literature review about SCI studies among the past two decades and then, some nuanced ideas about how to achieve superior SCI and how to make great use of high-level SCI for better performance are carefully studied.

Chapter 2 of this book has conducted a systematic literature review (SLR) on SCI to present an overview of prior studies, particularly the dimensions of SCI and its performance outcomes. SCI is widely accepted as a multi-dimensional construct, which consists of internal integration and external integration (including supplier integration and customer integration) (Flynn et al., 2010). Various methods are applied to conduct research on SCI, in which survey method appears most frequently. Scholars have employed multiple theoretical lenses to explain SCI issues. Resource-based view (RBV), organizational information processing theory (OIPT), contingency theory, and social capital theory (SCT) are ranked as the top four. These theoretical perspectives inspire researchers to examine the valuable functions, effects, antecedents, and contingencies of SCI. Nevertheless, SCI studies are still far from perfection. How some important emerging technologies will change the governance of supply chain partners and how SCI will influence the social and environmental performance of the firm, even the whole supply chain, are expected to be further investigated.

This chapter is co-authored by Xinyu Zhao and Yongyi Shou.

© The Author(s), under exclusive license to Springer Nature Singapore Pte Ltd. 2022   149
Y. Shou et al., *Supply Chain Integration for Sustainable Advantages*,
https://doi.org/10.1007/978-981-16-9332-8_8

150   8 Conclusions and Implications

In the five studies of this book, following the classic classification scheme of SCI, the relationships between SCI dimensions and the different roles of each dimension are investigated. Given the fast development of technologies and management practices, there are new research opportunities and gaps related to SCI. In this book, we have tried to fill some newly-identified gaps and offer state-of-the-art implications to practitioners.

8.2 Theoretical Implications

Studies in this book mainly contributes to the SCI literature. First, a number of antecedents are examined in our studies, most of which are largely ignored in previous research. At the product level, product complexity and variety are identified as important driving factors to influence the adoption of SCI (Shou et al., 2017). At the production system level, we explore the boundary conditions of SCI and find the contingency effects of internal production systems on external SCI–operational performance relationships (Shou et al., 2018). In other words, the performance effects of SCI depend on the internal production system configuration. Both studies remind us the importance to check the fit between SCI and the firm's internal factors, particularly the nature of its product and production system. In addition, we investigate the implementation of SCI from the socio-technical system (STS) perspective. Digital manufacturing technology (DMT) and human resource (HR) are typical technical and social factors, both of which significantly improve all three SCI dimensions and the effects of HR are moderated by competition (Tian et al., 2021). This study indicates the significance of the environmental subsystem in implementing SCI.

Second, this book sheds light on more possible links between SCI and firm performance, including those effects on risk management (Hu et al., 2020) and sustainability management practices (Kang et al., 2018), which have not received sufficient attention previously. SCI could not only be a direct driver of critical operational success, but also play an indirect complementary role of counteracting some adverse situations in the supply chain. To be more specific, SCI is testified as a great boundary-spanning intermediary between supply chain members with the moderating effects in a global supply chain risk management scenario. Since SCI enables better operations, better performance is accomplished.

In addition, this book extends the applications of theories and theoretical frameworks. Chapter 3 enriches the product–process fit literature by illustrating that such a fit needs to be realized among the entire supply chain network. Previous work on product–process fit focuses on the internal process within a firm while this study suggests that the process at the supply chain level also needs careful consideration to match the products, which are sourced, manufactured and delivered through the supply chain. Chapter 4 extends the application of OIPT and highlights the importance of matching information processing capacity with information processing needs in investigating SCI-related practices. In Chap. 5, how different subsystems of

8.2 Theoretical Implications

an STS will interact during SCI implementation is discussed. It is worth noting that the environmental subsystem plays an indispensable role in SCI.

Some emerging topics of the operations and supply chain management field are also touched in this book. Technologies have been a key driving force for business development, particularly in Industry 4.0. This book provides evidence that DMT plays a role in improving external SCI practices. Cloud technology may also boost supply chain information and physical integration (Bruque-Cámara et al., 2016). Whether and how digital technologies like blockchain will change the way of SCI is an interesting research question. Besides, supply chain risk management is becoming increasingly arresting. Risks will be very devastating when they affect supply chain members of different layers in cascade. Managing supply chain risks is an important topic and we believe the joint efforts across the supply chain will be an outstanding direction for future research. Lastly, sustainability concerns alongside supply chains ramp up rapidly. Manufacturers are eager to be greener and more prosocial, which exert pressures to their supply chain partners. One study in this book shows that the positive impact of both intra- and inter-organizational sustainability management practices (SMPs) on sustainability performance are strong. Hence, we call for more relevant discussion about supply chain level sustainability.

8.3 Managerial Implications

In the current globally networked economy, firms' facilities have been increasingly dispersed into diverse regions and countries to access valuable resources and meet customer needs. Manufacturers and their supply chains have undergone unprecedented increase of complexity. Meanwhile, fierce competition across global markets makes the effective supply chain management more and more important. This book provides abundant empirical evidence for the significance of SCI, including improving performance, supporting other strategic practices and so on. Managers should update their understanding about SCI and here we conclude some key ideas.

First of all, the successful implementation of SCI requires arduous work to provide various foundations and clear understanding about the contingency of the firm and industry.

Chapter 3 shows that firms should enhance their internal, supplier and customer integration when offering complex and diverse products. For a manufacturer offering highly complex products, information sharing and collaboration are needed across functional departments and supply chain partners in order to promote production coordination and problem solving (Kaufmann & Carter, 2006; Nickerson & Zenger, 2004). Similarly, a firm with high levels of product variety is expected to integrate within the organization and with external partners by means of information sharing, collaboration, joint decision-making and system coupling, in order to facilitate product development (Rothaermel et al., 2006) and to improve the production efficiency of suppliers (Al-Zu'bi & Tsinopoulos, 2012). Meanwhile, it is also suggested that product complexity/variety should be designed according to the current capability of SCI. If the level of a firm's SCI is relatively low, the firm may have poor capability in terms of the management of product complexity and variety. In short, product complexity/variety and the level of SCI should be well matched and balanced.

The results of Chap. 3 also suggest that internal and external integration should be treated as different practices. The firm should pay attention to the sequential ordering of the implementation of internal and external integration. More specifically, firms should give strategic priority to the implementation of internal integration in order to not only improve functional collaboration within the organization, but also facilitate external integration with their suppliers and customers. Particularly when the firm is characterized by a high degree of product complexity and variety, relationships with supply chain partners may involve exchange hazards and coordination difficulties. The implementation of external integration will be difficult unless the firm can build its capability through internal integration.

The impact of internal production systems is examined to understand the nuanced performance effects of SCI. Superior operational performance is critical for manufacturing firms to survive and thrive. Both researchers and practitioners regard SCI as a strategic means to enhance quality, flexibility, delivery and cost performance. However, "SCI is easier to talk about than to do" (Knemeyer & Fawcett, 2015) due to exchange hazards and coordination difficulties across organizational boundaries. Chapter 4 advances the existing understanding of when and how to implement an external SCI strategy according to the internal production system (i.e., one-of-a-kind production [OKP], batch production [BP] and mass production [MP] systems). The results indicate that SCI does not guarantee superior operational performance in all situations, as well as suggest that manufacturers should match external SCI with their internal production systems in order to achieve the desired operational performance. To be more specific, for firms equipped with an OKP-dominated system, there is a trade-off between customization and operational efficiency. The results of Chap. 4 suggests that OKP manufacturers should put more effort into customer integration, due to the high level of customization and technological complexity of the OKP system. A better strategy for OKP manufacturers to survive and grow is to strengthen customer integration for maximizing overall operational performance. However, BP manufacturers that pursue quality, flexibility and delivery performance can benefit from implementing supplier integration. By contrast, frequent information exchanges across organizational boundaries bring extra cost, such that manufacturers pursuing cost performance under BP-dominated conditions may not benefit from integration with suppliers and customers. Moreover, for firms equipped with an MP-dominated system, supplier integration plays a key role in enhancing cost performance. That said, neither supplier integration nor customer integration has a significant influence on other dimensions of operational performance. In other words, MP manufacturers need to strengthen supplier integration implementation for the maximization of cost performance since they have massive purchasing volumes from suppliers for the purpose of greater economies of scale.

Chapter 5 reminds managers that emerging enabling technologies like DMT are directly beneficial to SCI, as well as the HR practices. Competition also counts in

8.3 Managerial Implications 153

the HR–SCI relationship. As HR is more helpful in enhancing customer integration when competition is high, firms facing a high level of competition should invest in HR first, especially when confronted with resource limitations. Although both HR and DMT have positive effects on all the three dimensions of SCI, practitioners must pay close attention to how they prioritize these resources to build SCI efficiently and effectively in a specific context. When competition is low and firms maintain adequate resources, firms can invest either HR or DMT, or both, to foster SCI. However, when competition is high, we suggest the firm focusing on developing HR first. Chapter 5 provides some guidance about SCI implementation when facing intensive competition and resource limitation: firms trend to take customer-oriented strategies and likely to ignore internal issues, which means that internal integration will be affected.

Moreover, this book offers some new paths for managers to understand how SCI contributes to firm performance. In addition to the direct effects on firm performance, SCI also help firms to improve strategic practices.

Chapter 6 suggests that manufacturing multinational corporations (MNCs) can leverage external integration to counteract the weakened accuracy and timeliness of operations and supply chain information caused by broader international asset dispersion and further ensure the efficacy of supply chain risk management practices. Especially in markets with high complexity and uncertainty, coordination and collaboration with suppliers and customers will pay off with more valuable information. Therefore, supply chain managers of manufacturing MNCs with boarder international asset dispersion should emphasize the collaborative relationship with external supply chain partners given its imperative role in counteracting the negative influence caused by international asset dispersion. Given the complexity and challenges of global supply chains caused by COVID-19 and other global events, SCI deserves more attention from now on.

Chapter 7 further introduces the positive effects of SCI on sustainability. Sustainability has been a demanding goal of modern companies. Many governments have answered the call of the Paris Agreement. By building their SCI capabilities with internal and external supply chain partners, manufacturers can implement SMPs more efficiently and effectively to achieve their desired overall sustainability performance. Especially when firms operate their businesses under a high level of supply chain complexity, the relationship with supply chain partners can present collaboration and coordination difficulties (Bode & Wagner, 2015; Gimenez et al., 2012). In this situation, SMPs can become complicated and difficult to implement due to their connection with various supply chain partners. These challenges may be difficult to overcome unless manufacturers build up SCI capabilities that involve strategic collaboration, information sharing, joint decision-making and system coupling with supply chain partners. Given the increasingly high complexity of supply chains in the current business environment, well-designed and implemented SCI can facilitate information flow and close collaboration between supply chain partners, enabling manufacturers to successfully implement SMPs.

Different dimensions of SCI are also heterogeneous in building capabilities and enhancing performance. Intra- and inter-organizational SMPs require different types of SCI. Specifically, manufacturers need to build up all the three types of SCI to effectively implement intra-organizational SMPS. Alternatively, only the external SCI of suppliers and customers is necessary for inter-organizational SMPs. Considering that different types of SCI generate different effects on SMPs, manufacturers must clarify and classify the three types of SCI to align the appropriate type of SCI with better implementation of intra- and inter-organizational SMPs.

References

Al-Zu'bi, Z. M. F., & Tsinopoulos, C. (2012). Suppliers versus lead users: Examining their relative impact on product variety. *Journal of Product Innovation Management, 29*(4), 667–680.

Bode, C., & Wagner, S. M. (2015). Structural drivers of upstream supply chain complexity and the frequency of supply chain disruptions. *Journal of Operations Management, 36*(1), 215–228.

Bruque-Cámara, S., Moyano-Fuentes, J., & Maqueira-Marín, J. M. (2016). Supply chain integration through community cloud: Effects on operational performance. *Journal of Purchasing and Supply Management, 22*(2), 141–153.

Flynn, B. B., Huo, B., & Zhao, X. (2010). The impact of supply chain integration on performance: A contingency and configuration approach. *Journal of Operations Management, 28*(1), 58–71.

Gimenez, C., van der Vaart, T., & Pieter van Donk, D. (2012). Supply chain integration and performance: The moderating effect of supply complexity. *International Journal of Operations & Production Management, 32*(5), 583–610.

Hu, W., Shou, Y., Kang, M., & Park, Y. (2020). Risk management of manufacturing multinational corporations: The moderating effects of international asset dispersion and supply chain integration. *Supply Chain Management: an International Journal, 25*(1), 61–76.

Kang, M., Yang, M. G., Park, Y., & Huo, B. (2018). Supply chain integration and its impact on sustainability. *Industrial Management & Data Systems, 118*(9), 1749–1765.

Kaufmann, L., & Carter, C. R. (2006). International supply relationships and non-financial performance—A comparison of U.S. and German practices. *Journal of Operations Management, 24*(5), 653–675.

Knemeyer, A. M., & Fawcett, S. E. (2015). Supply chain design and integration: Why complex collaborative systems are easy to talk about but hard to do. *Journal of Business Logistics, 36*(3), 301–302.

Nickerson, J. A., & Zenger, T. R. (2004). A knowledge-based theory of the firm—The problem-solving perspective. *Organization Science, 15*(6), 617–632.

Rothaermel, F. T., Hitt, M. A., & Jobe, L. A. (2006). Balancing vertical integration and strategic outsourcing: Effects on product portfolio, product success, and firm performance. *Strategic Management Journal, 27*(11), 1033–1056.

Shou, Y., Li, Y., Park, Y., & Kang, M. (2017). The impact of product complexity and variety on supply chain integration. *International Journal of Physical Distribution & Logistics Management, 47*(4), 297–317.

Shou, Y., Li, Y., Park, Y., & Kang, M. (2018). Supply chain integration and operational performance: The contingency effects of production systems. *Journal of Purchasing and Supply Management, 24*(4), 352–360.

Tian, M., Huo, B., Park, Y., & Kang, M. (2021). Enablers of supply chain integration: A technology-organization-environment view. *Industrial Management & Data Systems, 121*(8), 1871–1895.

Appendix: Survey Questions

Construct	Item	Question
Indicate the current level of implementation of, action programs related to internal integration:		
Internal integration	II1	Sharing information with purchasing department (about sales forecast, production plans, production progress, and stock level)
	II2	Joint decision making with purchasing department (about sales forecast, production plans, and stock level)
	II3	Sharing information with sales department (about sales forecast, production plans, production progress, and stock level)
	II4	Joint decision making with sales department (about sales forecast, production plans, and stock level)
Indicate the current level of implementation of, action programs related to external integration:		
Supplier integration	SI1	Sharing information with key suppliers (about sales forecast, production plans, order tracking and tracing, delivery status, stock level)
	SI2	Developing collaborative approaches with key suppliers (e.g., supplier development, risk/revenue sharing, long-term agreements)
	SI3	Joint decision making with key suppliers (about product design/modifications, process design/modifications, quality improvement and cost control)

(continued)

© The Editor(s) (if applicable) and The Author(s), under exclusive license
to Springer Nature Singapore Pte Ltd. 2022
Y. Shou et al., *Supply Chain Integration for Sustainable Advantages*,
https://doi.org/10.1007/978-981-16-9332-8

155

156 Appendix: Survey Questions

(continued)

Construct	Item	Question
	SI4	System coupling with key customers (e.g., vendor-managed inventory, just-in-time, Kanban, continuous replenishment)
Customer integration	CI1	Sharing information with key customers (about sales forecast, production plans, order tracking and tracing, delivery status, stock level)
	CI2	Developing collaborative approaches with key customers (e.g., risk/revenue sharing, long-term agreements)
	CI3	System coupling with key customers (e.g., vendor-managed inventory, just-in-time, Kanban, continuous replenishment)
	CI4	Joint decision making with key customers (about product design/modifications, process design/modifications, quality)
Manufacturing network integration	MNI1	Improve information sharing for the coordination of the flow of goods between your plant and other plants of the network (e.g., through exchange information on inventories, deliveries, production plants, etc.)
	MNI2	Improve joint decision making to define production plans and allocate production in collaboration with other plants in the network (e.g., through shared procedures and shared forecasts)
	MNI3	Improve innovation sharing/joint innovation with other plants (through knowledge dissemination and exchange of employees inside the network)
	MNI4	Improve the use of technology to support communication with other plants of the network (e.g., ERP integration, shared databases, and social networks)
	MNI5	Developing a comprehensive network performance management system (e.g., based on cost, quality, speed, flexibility, innovation, and service level)

Indicate the current level of implementation of, action programs related to:

Digital manufacturing technology	DMT1	Use of advanced processes, such as laser and water cutting, 3D printing, and high-precision technologies

(continued)

Appendix: Survey Questions   157

(continued)

Construct	Item	Question
	DMT2	Development toward "the factory of the future" (e.g., smart/digital factory, adaptive manufacturing systems, scalable manufacturing)
	DMT3	Engaging in process automation programs (e.g., automated machine tools and handling/transportation equipment, robots)
Human resource	HR1	Delegation and knowledge of your workers (e.g., empowerment, training, encouraging solutions to work-related problems, pay for competence or incentives for improvement)
	HR2	Open communication between workers and managers (information sharing, encouraging bottom-up open communication, two-way communication flows)
	HR3	Workers' flexibility (e.g., multitasking, multi-skilling, job rotation)
Intra-organizational sustainability management	ISM1	Energy and water consumption reduction programs
	ISM2	Pollution emission reduction and waste recycling programs
	ISM3	Formal occupational health and safety management systems
Inter-organizational sustainability management	XSM1	Suppliers' sustainability performance assessment through formal evaluations, monitoring and auditing, using established guidelines and procedures
	XSM2	Training/education in sustainability issues for the suppliers' personnel
	XSM3	Joint efforts with suppliers to improve their sustainability performance
Supply chain risk management	SCRM1	Preventing operations risks (e.g. select a more reliable supplier, use clear safety procedures, preventive maintenance)
	SCRM2	Detecting operations risks (e.g. internal or supplier monitoring, inspection, tracking)
	SCRM3	Responding to operations risks (e.g. backup suppliers, extra capacity, alternative transportation modes)
	SCRM4	Recovering from operations risks (e.g. task forces, contingency plans, clear responsibility)

How would you describe the complexity of the dominant activity?

(continued)

(continued)

Construct	Item	Question
Product complexity	PC1	1—Modular product design; 5—Integrated product design
	PC2	1—Very few parts/materials, one-line bill of material; 5—Many parts/materials, complex bill of material
	PC3	1—Very few steps/operations required; 5—Many steps/operations required

Consider the importance of the following attributes to win orders from your major customers (1—Not important; 5—Very important)

Product variety	PV1	Wider product range
	PV2	Offer new products more frequently
	PV3	Offer products that are more innovative

How do you perceive the following characteristics of the environment in which your business unit operates?

Competition	CMP1	competitive rivalry: 1—Very low; 5—Very high
	CMP2	Market entry: 1—Closed to new players; 5—Open to new players
	CMP3	Threat that your products will become substituted: 1—Very low; 5—Very high
	CMP4	Bargaining power of suppliers: 1—Very weak; 5—Very strong
	CMP5	Bargaining power of customers: 1—Very weak; 5—Very strong

Please evaluate the probability of occurrence of the following risks:

Supply chain disruption risk	SCDR1	A key supplier fails to supply affecting your operations
	SCDR2	Your manufacturing operations are interrupted affecting your shipments
	SCDR3	Your shipment operations are interrupted affecting your deliveries

To what extent do you agree with the following statements?

Supply chain uncertainty	SCU1	Your demand fluctuates drastically from week to week
	SCU2	Your total manufacturing volume fluctuates drastically from week to week
	SCU3	The mix of products you produce changes considerably from week to week
	SCU4	Your supply requirements (volume and mix) vary drastically from week to week

(continued)

Appendix: Survey Questions

(continued)

Construct	Item	Question
How does your current performance compare with that of your main competitor(s)? *Relative to our main competitors, our performance is: 1—much lower; 3—equal; 5—much higher*		
Quality	Q1	Conformance quality
	Q2	Product quality and reliability
Delivery	D1	Delivery speed
	D2	Delivery reliability
Flexibility	F1	Volume flexibility
	F2	Mix flexibility
	F3	Product customization ability
Cost	C1	Unit manufacturing cost
	C2	Ordering costs
Time	T1	Manufacturing lead time
	T2	Procurement lead time
Relative to our main competitors, our performance is: 1—much higher; 3—equal; 5—much lower		
Social performance	SP1	Workers' motivation and satisfaction
	SP2	Health and safety conditions
Environmental performance	EP1	Materials, water and/or energy consumption
	EP2	Pollution emission and waste production levels
Please indicate your performance of the business unit in 2012. Compared to the three years ago the indicator is: 1—much lower; 5—much higher		
Business performance	BP1	Sales
	BP2	Return on sales (ROS)